Manual on the use of timber in coastal and river engineering

Matt Crossman
Jonathan Simm

Thomas Telford

HR Wallingford

Published by Thomas Telford Publishing, Thomas Telford Ltd, 1 Heron Quay, London E14 4JD.

URL: http://www.thomastelford.com

Distributors for Thomas Telford books are
USA: ASCE Press, 1801 Alexander Bell Drive, Reston, VA 20191-4400, USA
Japan: Maruzen Co. Ltd, Book Department, 3–10 Nihonbashi 2-chome, Chuo-ku, Tokyo 103
Australia: DA Books and Journals, 648 Whitehorse Road, Mitcham 3132, Victoria

First published 2004

Cover photograph
Pile extensions (courtesy Arun District Council)

A catalogue record for this book is available from the British Library

ISBN: 0 7277 3283 8

Typeset by HR Wallingford Ltd
Printed and bound in Great Britain by MPG Books, Bodmin, Cornwall

This report also constitutes the joint Defra/Environment Agency Flood & Coastal Erosion Risk Management Research & Development Programme Publication W5A – 069.

Preface

This document is intended to provide information and guidance on the use of timber in coastal, maritime and river engineering projects.

Timber has traditionally been used for the construction of a wide variety of engineering structures in or near water including groynes, jetties, lock gates and river bank protection. The fact that it is still used for many of these applications, despite advances in alternative materials is, at least partially, a reflection of the generally accepted characteristics that make timber an attractive choice of construction material. These are:

- renewable resource
- high strength to weight ratio
- high tolerance of impact and short duration loads
- good resistance to abrasion
- ease of construction, on site repairs and recycling
- attractive appearance
- natural durability (to a greater or lesser extent).

However, there are also a number of potential drawbacks including:

- inherent flaws and variability in properties associated with all natural materials
- limited availability of the very large section sizes, long lengths or high durability required for some engineering purposes
- species of timber native to the UK are only moderately resistant to the marine environment and are susceptible to biological hazards
- most species used in engineering applications are only renewable over a relatively long timescale, making it difficult to demonstrate sustainability, and few sources of tropical hardwoods are currently certified.

As a renewable resource, timber has the potential to be an environmentally responsible choice of material, particularly if recycled or obtained from sustainably managed forests. Although it is the only renewable construction material, negative publicity surrounding logging, particularly of tropical forests, has resulted in some reluctance from engineers to exploit the practical, environmental and aesthetic advantages of timber and there is an increasing reliance on alternative materials. Timber also needs different design and construction techniques from those used for materials such as concrete and steel. There has been some concern that the knowledge of experienced engineers is being lost as they retire and the skills required to use timber efficiently and effectively are not being passed on to younger engineers.

This manual addresses these issues by providing an introduction to the use and properties of timber, a framework for responsible selection and procurement of timber materials (including recycled timber), and a discussion of design and construction issues. Different types of timber structures are documented, and management and maintenance practices are described.

The manual is intended for practising engineers and may also be a useful reference for students, procurement specialists and managers wishing to understand more about the use of timber in coastal and river environments.

Acknowledgements

This manual has been prepared as the primary deliverable from an HR Wallingford research project carried out under the Department of Trade and Industry 'Partners in Innovation' scheme. (N.B. The Contract was started under the Department of the Environment, Transport and the Regions, and transferred to the Department of Trade and Industry during financial year 2001/2). The objective of the research was to collate existing knowledge, experience and research in order to develop best practice guidance on the use of timber in coastal and river engineering.

The manual was prepared by HR Wallingford and published on behalf of the Department of Trade and Industry and the Environment Agency. The views and information presented are those of HR Wallingford and whilst they reflect the consensus of the Steering Group, they are not necessarily those of the funding organisations. The project was managed by Matt Crossman under the guidance of Jonathan Simm of HR Wallingford.

The manual was edited by Matt Crossman and Jonathan Simm from text prepared by John Williams, Simon Howard, John Roach, Rod Nelson, Jeremy Purseglove, Brian Holland, Udo Perdok and David Harlow. Additional editorial support was provided by Clive Orbell-Durrant and Celia Yardley.

Further information on some of the issues explored within this manual is provided on the associated website at: www.timbermanual.org

Project funders
Cash funding for the project was provided by the following organisations:

- Department of Trade and Industry (Partners in Innovation programme)
- Environment Agency
- HR Wallingford
- TRADA Technology
- British Waterways
- SCOPAC.

In kind contributions
In kind contributions, including data, staff time and other resources committed to the project were provided by the following organisations:

- Aitken & Howard Ltd
- Arun District Council
- Babtie Group
- Bournemouth Borough Council
- British Waterways
- Dean and Dyball
- Ecosylva
- Environment Agency

- The Institution of Civil Engineers
- John Martin Construction
- Mackley Construction
- Mott MacDonald
- Mowlem
- Posford Haskoning
- SCOPAC
- Shoreham Port

- GR Wiltshire & Co.
- HR Wallingford
- TRADA Technology
- WS Atkins

Steering Group

The research was guided by a steering group comprising the following members:

John Andrews	Posford Haskoning
Peter Anidjar-Romain	WS Atkins
Rob Bentinck	Mowlem Marine
David Bligh	British Waterways
Andrew Bradbury	SCOPAC
Mark Buttle	GR Wiltshire & Co.
Tony Camilleri	Mackley Construction
Richard Copas	Environment Agency
Matt Crossman	HR Wallingford
David Harlow	Bournemouth Borough Council
Mike Hodgson	John Martin Construction
Brian Holland	Arun District Council
Simon Howard	Posford Haskoning
Paul Kemp	Aitken & Howard Ltd
John Laker	Dean & Dyball Construction
Rod Nelson	Ecosylva
Clive Orbell-Durrant	Independent Consultant
Tony Parker	Posford Haskoning
Jeremy Purseglove	Mott MacDonald
John Roach	Babtie Group
Jonathan Simm (Chairman)	HR Wallingford
Tony Vaughan	Shoreham Port Authority
Brian Waters	Institution of Civil Engineers
John Williams	TRADA Technology

Assistance was also received from:

Martin Luker	Environment Agency
Stephen McFarland	Canterbury City Council
Stephen Cook	New Forest District Council
Maaike van der Kroon	HAS Den Bosch
Udo Perdok	TU-Delft
Henk Jan Verhagen	TU-Delft
Wim Bak	Rijkswaterstaat

HR Wallingford are grateful to the funders, the Steering Group and all those who have contributed time and information to the project. Particular thanks are due to Udo Perdok who made a substantial contribution to the success of the project whilst at Wallingford as part of his MSc at Delft University of Technology. We would also like to thank Maaike van der Kroon for her input to the environmental issues, Bev Reader for her organisational and administrative assistance and those involved in the finalisation of the document, including

Clive Orbett-Durrent, Celia Kirby, Mike Wallis and the staff of HR Wallingford's Report Production Unit.

HR Wallingford carries out advanced research, consultancy and software development relating to civil engineering hydraulics and the water environment. Established in 1947 as a Government research centre, it is now an independent not-for-profit company employing over 200 engineers, scientists, mathematicians and support staff. HR Wallingford is concerned with all aspects of water management and engineering in catchments, rivers, estuaries, coasts and offshore. It carries out predictive physical and computational model studies, desk studies and field data collection backed by large-scale laboratory research and the development of advanced computer simulation technology.

HR Wallingford, Howbery Park, Wallingford, OX10 8BA, UK
www.hrwallingford.co.uk
Telephone: 01491 835381
Contact email: j.simm@hrwallingford.co.uk

Glossary

Accreditation body	An independent authority which examines the operations and capacity of certification bodies
Baulk	Piece of square sawn or hewn timber of equal or approximately equal cross-section dimensions of size greater than 100 mm x 125 mm
Bole	The stem, or trunk of a tree when over 200 mm diameter
Bone	Gradient, slope of a section of groyne
Boxed heart	A log converted so that the centre of the heartwood (perishable pith) is wholly contained in one piece and surrounded by durable heartwood
Breastwork	Vertical or steeply sloping structure constructed parallel to the shoreline or bank at or near the crest to resist erosion or flooding
Certificate	A document that attests that a process or a product meets a standard
Certification body	An independent body that conducts inspection and verification, and which issues certificates (of conformity to a standard)
Chain of custody	The distribution channels between forest and end user
Check	Separation of fibres along the grain forming a crack or fissure that does not extend through the timber or veneer from one surface to the other
CITES	Convention on International Trade in Endangered Species of Wild Fauna and Flora
Contiguous pile	Term usually applied to cast-in-place concrete piles immediately adjacent to or touching each other. Sometimes used for Plank Piles (see below)
Finger joint	Pieces of timber end jointed by glued interlacing wedge-shaped projections
Forestry stakeholder	A stakeholder is 'an individual or organisation with a legitimate interest in the goods and services provided by a forest management unit' – forestry stakeholders are likely to include government representatives, forest scientists, forestry protection organisations, indigenous forest dwelling people and/or local people and their representatives, forest managers and the timber trade

FSC	Forestry Stewardship Council
Green timber	Timber freshly felled or still containing free moisture that is not bound to the cell wall
Heartwood	Inner zone of wood that, in a growing tree, has ceased to contain living cells and reserve materials
Hewn	Timber section squared or levelled with an axe or adze
Intermediate wood	See transition wood
King pile or post	A pile or post acting in cantilever to support a retaining structure
Knee pile	(King) pile at the change of 'bone' in a groyne
Knot	Portion of a branch embedded in the wood resulting in a discontinuity in the grain
NGO	Non-governmental organisation
Plank pile	Individual planks driven (normally vertically) adjacent to each other to form large panels, often with the driving end cut at an angle or shod with metal for ease of driving and to ensure that adjacent planks provide a tight fit
Planking	Long flat timber boards (horizontal or sloping) attached to a frame, for example forming a groyne or deck
Revetment	Protective structure parallel to a bank, cliff or slope. Non load-bearing
River Wall	A structure which both protects and supports a river bank, either vertical or steeply sloping
Sapwood	Outer zone of wood that, in a growing tree, contains living cells and food reserve materials (e.g. starch), generally lighter in colour than heartwood though not always clearly differentiated
Shake	Separation of wood fibres along the grain irrespective of the extent of penetration
Sheeters	Infill boarding between posts, piles, etc. in any orientation
Sheet pile	Steel sheet piles of any section. Sometimes used for plank piles
Slope of grain	Angle between the direction of the grain and the axis of the piece (the greater the angle, the greater the strength-reducing effect caused by slope of grain)

Split	Separation of fibres along the grain forming a crack or fissure that extends through timber
Standards	Documented agreements containing technical specification of products or processes
Straight grain	Grain that is straight and parallel or nearly parallel to the longitudinal axis
Transition wood	Found in a few species of tropical hardwood and sandwiched between the sapwood and heartwood. Transition wood is not as durable as heartwood, and its presence may be a contributory factor to the apparent failure/erosion of timber components
Transom	An intermediate horizontal timber beam acting as a strengthening or supporting member in a structure
TTF	Timber Trades Federation
Waling	A horizontal or sloping beam which distributes loading between the piles of a groyne or breastwork and provides a point of attachment for plank piles
Wane	Term applied to converted wood in which the corner is missing at the circumference of the log, due to economical conversion

Contents

Illustrations

Boxes

1. Introduction

1.1. BACKGROUND AND PURPOSE OF THE MANUAL

This manual was conceived to fill a major gap in the practical guidance regarding the use of timber available to coastal and river engineers. Many of the standard references that deal with timber are limited to the use of softwoods for buildings, and few civil engineering courses at universities provide a thorough introduction to the use of timber. With increasing pressure on engineers to provide environmental benefits within schemes whilst ensuring that works are sustainable as well as technically and economically sound, there is considerable scope for the increased use of timber.

This manual has been prepared therefore to facilitate the efficient and effective use of timber structures in and around water. It concentrates on the present best practice in selecting, designing, specifying, procuring, constructing and managing timber structures. Some information is provided on details for specific structures, but the manual is intended to complement rather than replace established design references, national or international standards and codes of practice.

It is also hoped that this manual will contribute to improved design and construction practices with timber. It is intended to encourage more durable structures with their resulting cost savings and reducing timber wastage. The use of timber needs to be set in the context of other available new and recycled materials. This aspect is addressed briefly (in Chapter 2), but for a fuller discussion of the environmental issues associated with the selection of materials the reader is directed to Masters (2001).

1.2. READERSHIP AND USE OF THE MANUAL

The manual is principally intended for practising engineers with some knowledge and experience of coastal or river engineering who wish to review advice or best practice relating to a specific issue or structure. However, it will also provide students, recent graduates and other professionals (including clients, environmental or procurement specialists) with an overview of the various issues associated with the use of timber in coastal and river engineering. It is likely that the former will find the manual of most use as a reference source, reviewing specific sections as and when issues arise, whilst the latter may benefit from following the text in a more ordered fashion.

It should be noted that a document such as this can never be a substitute for the judgement and understanding of a qualified and experienced engineer. It is hoped that the manual will provide a useful introduction and contribute to improved practice in the selection, design, specification, construction and maintenance of timber structures, but it cannot possibly cover the full range of issues and considerations in such a complex environment. It is strongly suggested that the services of capable and experienced engineers and/or consultants with specialist timber knowedge are employed if any further information or guidance is required.

1.3. STRUCTURE OF THE MANUAL

The manual (see flowchart in Figure 1.1) is structured to follow the same process as most engineering projects. Chapter 2 provides an introduction to the use of timber in coastal and river engineering, including information required at scheme development or feasibility stages, such as typical life cycles and alternative materials. Chapter 3 provides an introduction to the properties of the raw material – wood – while Chapter 4 describes the processing of wood into timber, its properties and ways in which these can be improved, including the use of manufactured sections and preservative treatment.

Chapter 5 provides information on the environmental issues associated with the production and use of timber and sets out a pragmatic framework for the responsible selection and procurement of timbers. It is hoped that this will encourage and reward further moves towards sustainable forest management, whilst maintaining the flexibility for timbers with appropriate properties to continue to be used in coastal and river engineering in the interim period.

The various stages in the design process are described within Chapter 6, with each stage discussed in some detail. General design and construction issues are discussed in Chapter 7. Chapter 8 provides information and example illustrations of the wide range of timber structures used in coastal and river engineering, with references to useful sources of design information.

Chapter 9 provides information on the monitoring and assessment of timber structures, including a description of how timber degrades, methods of inspecting timber structures and identifying timber species. The description of the life cycle of timber structures is completed in Chapter 10 which describes the maintenance, repair and adaptation activities that are vital to the continued effectiveness of many timber structures.

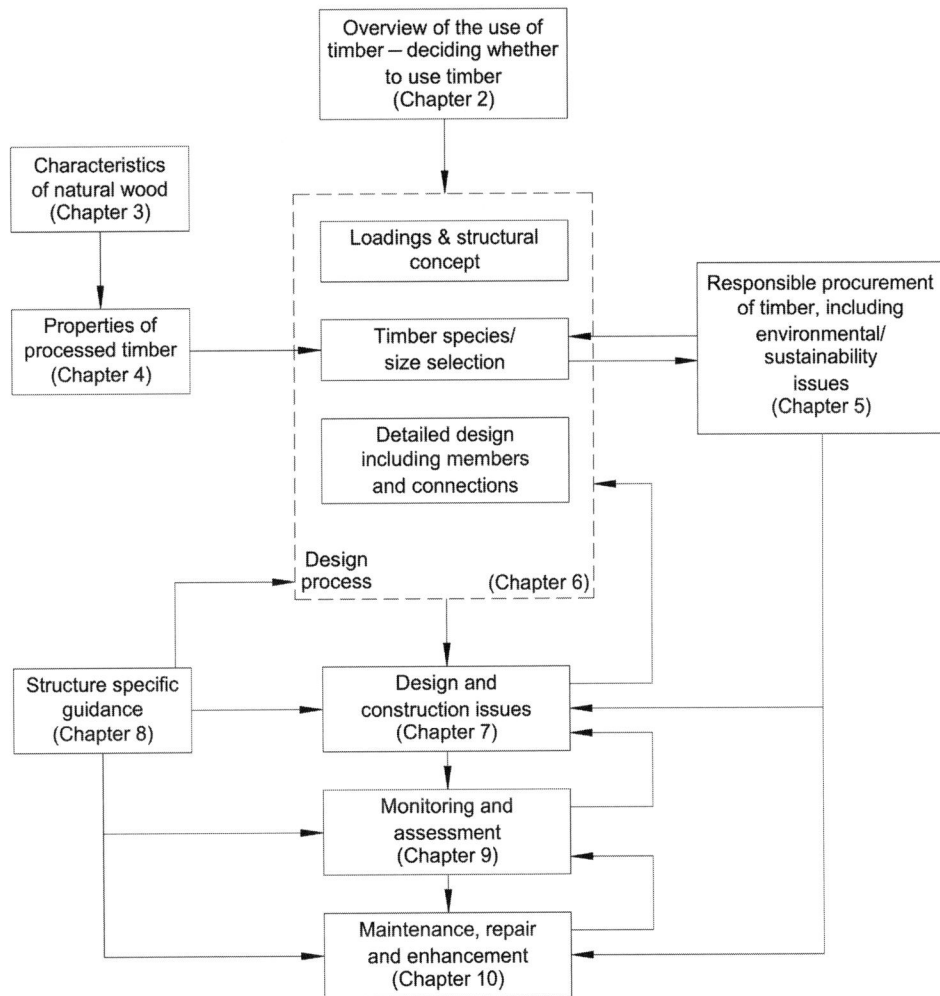

Figure 1.1. Flowchart illustrating concept for manual

2. Overview of the use of timber in coastal and river engineering

Timber has been used as a material for hydraulic engineering structures for centuries. It is used in groynes to retain and control beaches, revetments and bank protection as well as piers, jetties, lock gates and navigation structures. Although alternatives such as concrete, steel and rock have become more popular, timber can still offer advantages of sustainability, cost and appearance.

2.1. WHY TIMBER?

Timber has a combination of properties, which make it a very attractive choice of construction material for coastal and river environments. These properties include:

- renewable and environmentally responsible
- high strength to weight ratio
- good workability and ease of modification, repair or reuse
- natural durability
- high tolerance of short duration (shock) loads
- attractive appearance.

Timber structures are located in many of our most valued and precious environments, and the suitability of timber for various functions is well demonstrated by its use in both traditional and contemporary structures. One of the most significant uses of timber in coastal and river engineering is for groynes to control beaches which have featured in seaside landscapes for generations. Many of these groynes, including those at Eastbourne (Figure 2.1) are aesthetically pleasing as well as being efficient, effective and environmentally responsible.

Figure 2.1. Timber groynes at Eastbourne, East Sussex (courtesy Posford Haskoning)

2.2. ENVIRONMENTAL ISSUES

Timber is an environmentally responsible material option if recycled or obtained from a sustainably managed resource. It is particularly attractive since it has low embodied energy and is virtually 'carbon neutral' (apart from transport and processing). In addition, all forests sequester carbon from the atmosphere both in living biomass and in forest soils. Living forests (Figure 2.2) also have important functions in the regulation of other cyclical processes important to the earth's climate – particularly air quality and the water cycle.

Chapter 5 provides a detailed description of the environmental issues and presents a framework for the responsible use of timber. It is important to note that whilst there is concern regarding the loss and degradation of forests, most types of forest can be managed sustainably, that is to say harvested periodically for timber and other products in such a way that their productive benefits for future generations are not compromised. Indeed, it is often argued that the production of timber provides an economic value to forests which contributes significantly to their protection. Issues such as illegal logging, deforestation and wastage must be addressed, but it should be remembered that many alternative materials are significantly more damaging to the environment than responsibly sourced timber.

Figure 2.2. Living forest (courtesy Timber Trades Federation)

2.3. LIFE CYCLE ANALYSIS OF TIMBER STRUCTURES

To ensure that a scheme is sound, sustainable and appropriate the process of developing, selecting and optimising individual structures and the scheme as a whole is informed by analysis of the wide range of environmental, economic and technical issues over the whole life of the scheme.

2.3.1. Whole life costs

The series of costs over the life cycle of any particular scheme can be expressed as a Present Value using standard accounting techniques, including discounting future expenditure, to reflect the expected return on capital. Depending on the assumed discount rate, this can have a significant impact on the types of works which are most attractive. The choice of schemes may also be influenced by the availability of grants or other funds for capital or maintenance works. The issues associated with whole life costs are dealt with in detail by Simm and Masters (2003) but it should be noted that such analysis should include all of the following costs:

- research, analysis and design
- capital works

- monitoring and periodic review
- maintenance works
- termination (including decommissioning and disposal or reuse of materials)
- risk (including the accuracy of cost predictions and disruption or damages resulting from failure).

Whilst capital costs can often be determined relatively accurately, the costs associated with monitoring, maintenance and termination are often much less predictable. Nonetheless, for publicly funded projects in the UK, there is currently a Treasury requirement to assess projects over a 100 year time period using discount rates varying between 3.5% and 2.5%. The way in which the more uncertain time–distant costs are assessed may influence the scheme selection process, even though the least predictable elements of the costs are often those taking place furthest into the future and as such are most heavily discounted.

2.3.2. Environmental impacts

The sustainable use of new and recycled materials in coastal and river construction is described by Masters (2001) which includes the following hierarchy of materials sourcing options:

1. suitable materials available on-site from a previous scheme or structure
2. locally-sourced reclaimed or recycled materials appropriate to fulfil functional requirements
3. recycled or reclaimed materials from further afield that can be delivered to site, predominantly by sea or rail, or locally sourced primary materials
4. reclaimed or recycled materials transported from further afield, predominantly by sea or rail
5. primary materials transported from further afield by road.

Whilst this presupposes the building of a new structure it should be noted that, where it is possible, the modification, repair or maintenance of existing structures may result in less environmental impacts. The hierarchy identifies the importance of using reclaimed or recycled materials wherever practical and also highlights the significance of transport in determining the selection of materials. The Ecopoints estimator spreadsheet tool developed for the manual enables an objective analysis of scheme options comprising different materials. It provides a quantitative assessment of the wider environmental impacts, material and transport options (100 Ecopoints is equivalent to the total impacts of a single UK citizen over 1 year), which can be used alongside cost estimates to inform the selection of the preferred scheme as shown in Box 2.1.

2.3.3. Engineering issues

Whilst useful, the environmental and economic analyses cannot be used as the only basis for making a decision. There can be significant differences in the performance of the different options and the selection of a particular scheme must also consider the practicality and reliability of the various components and the scheme as a whole (see Section 2.5) as well as local environmental issues (described in Section 7.2).

Box 2.1. Comparison of rock and timber groynes

Rock and timber groynes have very different characteristics. Timber groynes are vertical structures and use only relatively small quantities of material but due to the hostile environment and potential for biological attack, tropical hardwoods are commonly selected and transport distances are considerable. Rock groynes use much greater volumes of materials but the transport distances are often less. Since construction costs, performance and environmental impacts are highly dependent on the location and function of the structure, this analysis is only applicable to a particular location on the south coast of England.

Cost

The unit cost (per metre length) of different types of groyne was calculated for a scheme appraisal period of 50 years. This includes monitoring and maintenance costs for both types of groyne, but is dependent on the discount rate used. The analysis suggests that for low groynes rock is cheaper, but that where significant height is required (for example, near the top of the beach) timber is likely to be less expensive.

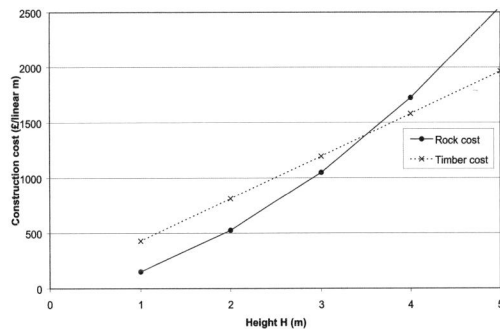

Environmental issues - Ecopoints

The Ecopoints analysis uses material quantities from the construction cost estimate, combined with transport distances and methods for each material. It does not include local environmental impacts (such as the provision of habitat in rock structures) or the actual construction activities, but the results suggest that the timber structure will have less impact in many situations.

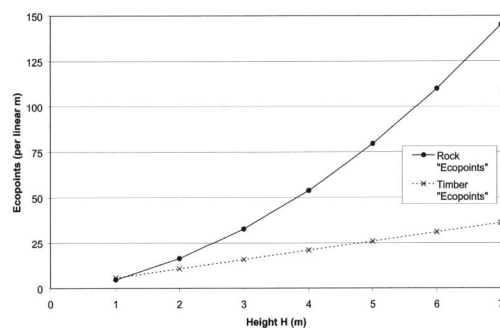

2.3.4. Typical life cycles for timber structures

The performance and service life of timber structures are affected by a wide range of issues including:

- severity of exposure (e.g. biological and/or mechanical damage as well as structural loading)
- durability and quality of materials (fastenings and fixings as well as timber components)
- quality of design, detailing and construction (particularly joints and connections)
- frequency of monitoring and early implementation of maintenance or remedial works.

In practice, timber structures most often fail through a process of gradual deterioration of the structure to the point where it no longer functions effectively and, if dangerous for the general public, may require demolition. In many situations the service life of durable hardwood (e.g. greenheart) elements is of the order of 25 years, however a much longer service life has been experienced and in some instances timber over a hundred years old has been reclaimed from redundant structures for reuse in new structures.

At the design stage, expectations of service life are based generally on experience of timbers with similar properties employed in similar conditions. Where such information is not available, an indication of possible service life may be obtained from a review of the likely exposure to biological attack, abrasion or other loadings. An understanding of the relative performance and life cycles of different timbers in particular situations may be obtained from laboratory testing under accelerated laboratory conditions (see Box 4.6) but to date trials have always been conducted in prototype on working structures.

2.4. ALTERNATIVE MATERIALS

The attractiveness, with respect to wider environmental impacts, of timber groynes over rock groynes in many situations is demonstrated in Box 2.1. However, this is at least partially attributable to the difference in structural form of the two types of groynes and it is often possible to substitute alternative materials for timber whilst retaining a similar structural form. Whilst the technical issues associated with the use of alternative materials will vary considerably with the type of structure and particular situation, a simple analysis of piles for conventional groynes has been undertaken and is described in Box 2.2.

Box 2.2.   Comparison of timber with other structural materials

Long lengths of tropical hardwoods are used traditionally for the piles of conventional timber groynes (on the south coast of England piles are frequently constructed from greenheart with sides of 230 or 305 mm). To assess alternative materials, dimensions for components providing similar structural properties have been determined and combined with likely transport details and the anticipated service life to quantify wider environmental impacts using the Ecopoints estimator (Masters 2001):

Material	Nominal dimensions (mm)	Min.expected life (years)	Transport distances (km)		Ecopoints (per pile m)
			Ship	Lorry	
Tropical hardwood	230 × 230	25	6500	200	0·8
Recycled hardwood	230 × 230	10 / 20	–	100	0.5 / 0.1
Pitch pine	285 × 285	15	5500	200	2.1
Oak	305 × 305	10	100	300	0.8
Douglas fir	315 × 315	7	–	500	1.9
Plastic (reinforced HDPE)	250 dia	50	5500	200	3.1
Reinforced concrete	375 × 375	30	–	300	1.4
Steel – Universal column	254×254×86	30	–	300	3.5

The results demonstrate that the use of recycled timber is the most favourable, providing it has a reasonable service life and the transport distances are not excessive. Where appropriate recycled timber is not available the use of either tropical or temperate new hardwoods is preferable. The Ecopoints estimator is not sufficiently precise to enable a distinction to be made between the two types of hardwood and in any case such an assessment would probably require a detailed review of the sustainability of forest management practices, transport methods and processing as well as the exposure conditions. In practice, other considerations such as the availability of large section sizes and cost/impacts of construction activities are likely to favour the use of tropical hardwoods. The use of precast reinforced concrete elements appears to be more favourable than softwoods, although this could be reversed if the service life for concrete is reduced dramatically due to abrasion. Finally, plastic and steel are least favoured, although it should be noted that there is no facility within the Ecopoints estimator to consider the use of recycled plastic or the plastic–timber composites which are being developed in the USA. Consideration of these materials would require further information on source materials and manufacturing processes as well as alternative uses (such as direct reuse, recycling for alternative products or energy recovery).

More extensive information (including case studies) on the use of different materials in coastal and river engineering is provided by Masters (2001). Whilst it is apparent that hardwoods are favoured in many situations, there are some specific circumstances where the use of timber is not practical. These include locations where biological attack is particularly severe (such as tropical areas or structures in the vicinity of cooling water outlets), access is difficult or hazardous and where strengths, lengths or sections sizes in excess of those available for timber are required.

2.5. ENGINEERING TIMBER STRUCTURES

The design and construction of timber structures is markedly different from the more widely used steel and concrete with which many engineers are familiar. In addition, most of the timber used in normal or domestic construction does not need to have particularly good durability or strength, and many of the standard references on structural timber design thus concentrate on softwoods or manufactured softwood sections. However, timber employed in coastal and river engineering is often subjected to large loadings and harsh environmental conditions, and it is no surprise therefore that the durability, strength and large section sizes afforded by tropical hardwoods are particularly valued in this field.

As a natural material, timber is inherently variable, with considerable differences in properties between different species making it important that appropriate timbers are selected for use in any given situation. The design and construction of structures in coastal and river environments requires particular care in the selection of appropriate materials, which is further complicated by the need to ensure that such materials are environmentally responsible. Engineers have traditionally used a limited number of timber species for which considerable experience is available, but there is increasing pressure to make more use of lesser known species and new testing techniques (for example, the work described in Box 4.6) should enable the performance of these species to be predicted more accurately.

The life cycle for coastal and river timber structures involves seven stages as illustrated in Figure 2.3. There is a need for continuing engineering through each of these stages of the asset life. Information and experience from each stage feeds into subsequent activities enabling the structures to evolve over time. The benefits and disadvantages associated with different proportions of capital and maintenance works can also influence scheme choice. Typically, timber elements have relatively short useful lives and structures have significant monitoring and maintenance obligations. This may not be practical where access is difficult or dangerous and is often cited as a disadvantage. However, it can be argued that the need to maintain or replace individual structures provides opportunities for modification or adaptation during the scheme life, and in some circumstances (such as a large groyne field or network of lock gates) the individual structures can be replaced on a rolling programme of approximately the same duration as the structure life (Box 2.3). This enables expenditure to be maintained at a relatively steady level whilst also facilitating a long term relationship with external contractors, evolution of design and construction practices, and continuity of knowledge and experience.

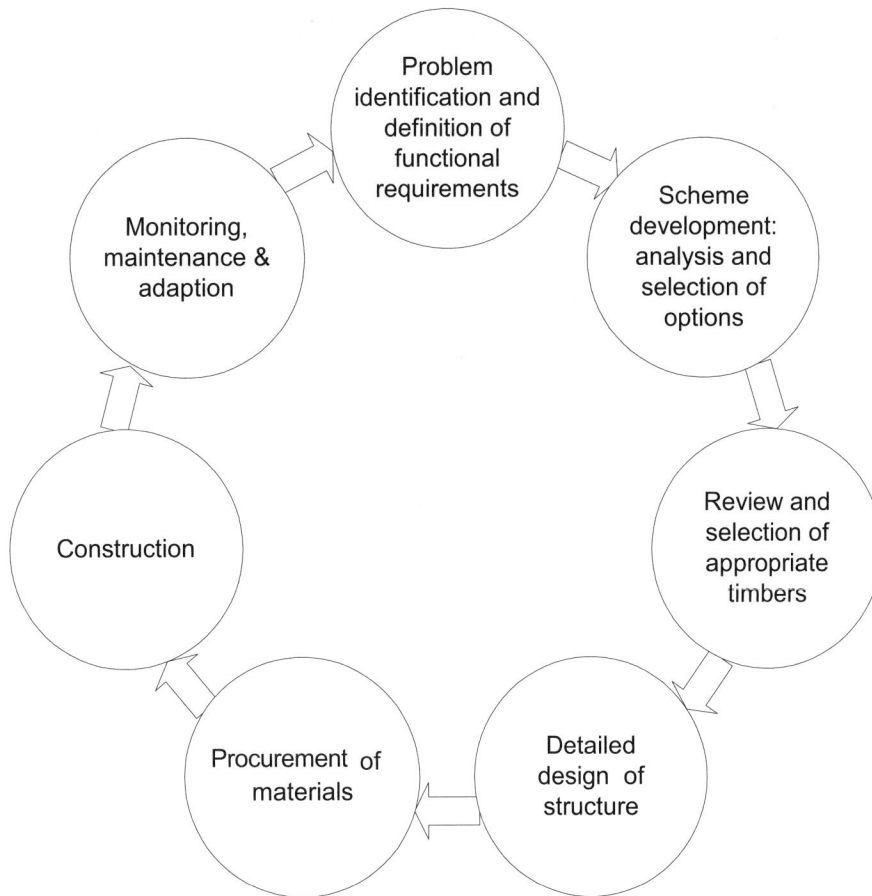

Figure 2.3. Typical life cycle for timber structures

Box 2.3. Bournemouth Borough Council rolling groyne reconstruction programme

The Bournemouth Borough Council maintain and replace 51 timber groynes along their coast. The groynes are constructed from tropical hardwoods to retain a sand beach and have a life expectancy of approximately 25 years. A rolling programme has been established to replace two groynes each year. This has the following advantages:

- continuity in design and construction experience
- opportunity to incorporate evolutionary refinements in groyne construction as they are developed
- opportunity to adjust groyne profiles to changing beach profiles and rising sea levels
- opportunity to refine groyne spacing.

14

Box 2.3.   Bournemouth Borough Council rolling groyne reconstruction programme (continued)

The present construction cost of each groyne is approximately £200 000 which is, in part, due to the small tidal range and length of the groynes necessitating considerable temporary works (a steel platform is usually constructed to provide access for a crane). The hard underlying strata necessitate pre-boring for the piles with high-pressure water lances. Monitoring beach levels within the groyne bays assesses the performance of the groynes and has led to several improvements in both the profiles and positions of the groynes. The groynes themselves are also regularly inspected and prioritised for replacement.

The extensive use of timber is a testament to its versatility and ease of use; examples and illustrations of a wide range of timber structures are provided in Chapter 8, along with specific design details which have proven to be reliable and effective. The challenge for the current generation of engineers, scientists and timber specialists is to continue the processes of innovation and refinement, ensuring that we continue to make best use of valuable resources and maintain the competitiveness of timber structures.

3. The natural characteristics of wood – a brief introduction

An understanding of the characteristics and properties of *wood* as a natural raw material will enable the designer or user to ensure that the *timber* produced from it is used to best effect. Unlike the many manufactured materials used in engineering, efficient utilisation is dependent on some form of selection and grading. However, timber has the advantage over almost all other materials in coming from a living, renewable resource. Good land management and judicious felling regimes are recognised as essential facets of the timber trade that will help to secure the long term availability of certain timbers and also to promote timber originating from well managed forest resources. The need for, and implications of, sustainable forest management and certification are discussed in Chapter 5.

3.1. THE STRUCTURE OF WOOD

Wood is made of organic matter. The basic building block is the wood microfibril, which may be described as a fibre composite where the fibre element provides tensile strength to the composite while the matrix provides stiffness and transfers stress from fibre to fibre. The microfibril comprises cellulose, hemicellulose and lignin. The fibre constituent is made up of cellulose, which provides strength, and the hemicellulose and lignin act as the matrix that stiffens and bonds the cellulose fibres.

Cellulose and hemicellulose are sugar-based polymeric 'building blocks'. Cellulose comprises building blocks of glucose that are linked up longitudinally to form long, thin filaments that lie parallel to each other in a particular pattern giving the cellulose component a high degree of crystallinity. A single molecule of cellulose is made up of a chain of approximately 8000 glucose units. It is this arrangement of the glucose building blocks that imparts strength to cellulose.

Hemicellulose is similar to cellulose in that it is made up of various carbohydrate units (sugars) such as mannose and galactose but is not as ordered in its structure and is described as having a low degree of crystallinity. A typical hemicellulose molecule may contain a chain of 150-200 sugar units.

Lignin is a complex non-crystalline compound comprising many different organic constituents and may be summarised best as a matrix of aromatic molecular compounds.

Other chemicals may be present in the wood. These chemicals may be classified as extractives. Examples of extractives are gums, oils, tannins, latex, resins, silica and calcium deposits. Large quantities of silica may cause blunting of cutting tools but are also thought to be responsible for imparting greater resistance against attack by marine borers. To summarise, the make-up and distribution of extractives vary from species to species and are thought to play a pivotal role in imparting durability against biological attack. Durability is discussed in greater detail in Section 4.5.

All living organisms are composed of cells. In the living tree, different cell tissues perform different tasks. Some tissue groups convey water and nutrients, and others perform a structural function providing the tree with strength and elasticity.

Most of the conducting and supporting tissue is arranged vertically and this arrangement forms the grain of the timber. Of course, water and nutrients have to be transported horizontally, across the grain, as well. This is carried out by an arrangement of horizontal tissue types known as rays. The size and distribution of the ray tissue vary from species to species and are a useful diagnostic feature for identification. Figure 3.1 illustrates the typical gross features of timber.

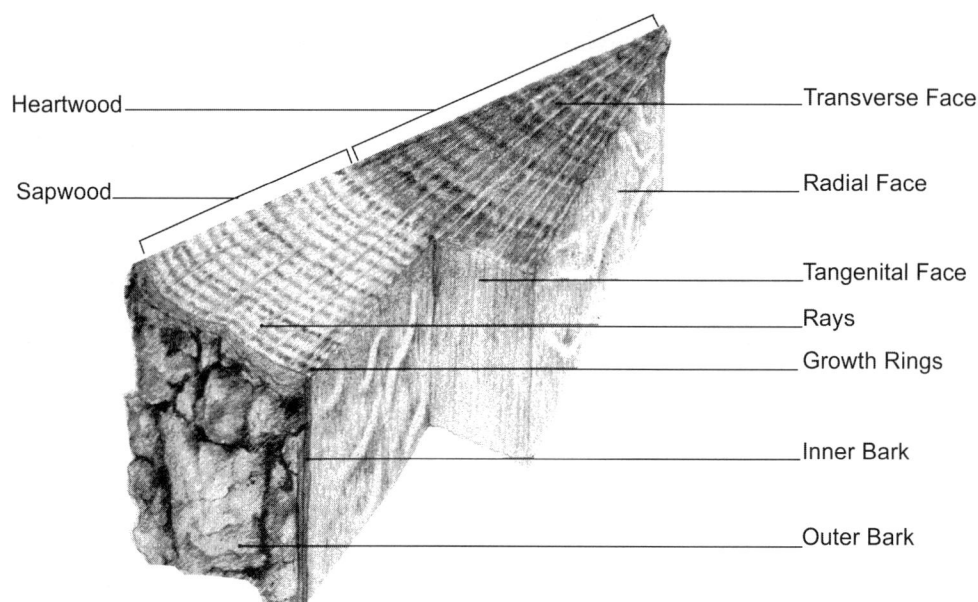

Figure 3.1. Gross features of wood (courtesy TRADA Technology)

3.2. CROSS-SECTIONAL FEATURES

For every year of growth, the tree will lay down a ring of timber known as an annular ring. This emerges from the cambium, a thin sheath of cells between the bark and wood from which all cells originate. The outermost ring is the most recently formed. This ring may be further broken down into earlywood and latewood. Simply put, earlywood is laid down during the growing season where transportation of nutrients and water is of primary importance and latewood is laid down at the end of the growing season where metabolic activity is at its lowest level. The primary function of latewood is to provide support.

The cross-section of the trunk of the tree may be divided into two zones: the sapwood and the heartwood (Figure 3.2). The sapwood is physiologically active and supports the tree's metabolism, i.e. living activity. The sapwood is usually a narrow, paler band of timber encompassing the heartwood. However, not all trees show a clear difference between sapwood and heartwood, e.g. spruce and greenheart. In all cases, the sapwood should be viewed as having low resistance to all types of biological attack.

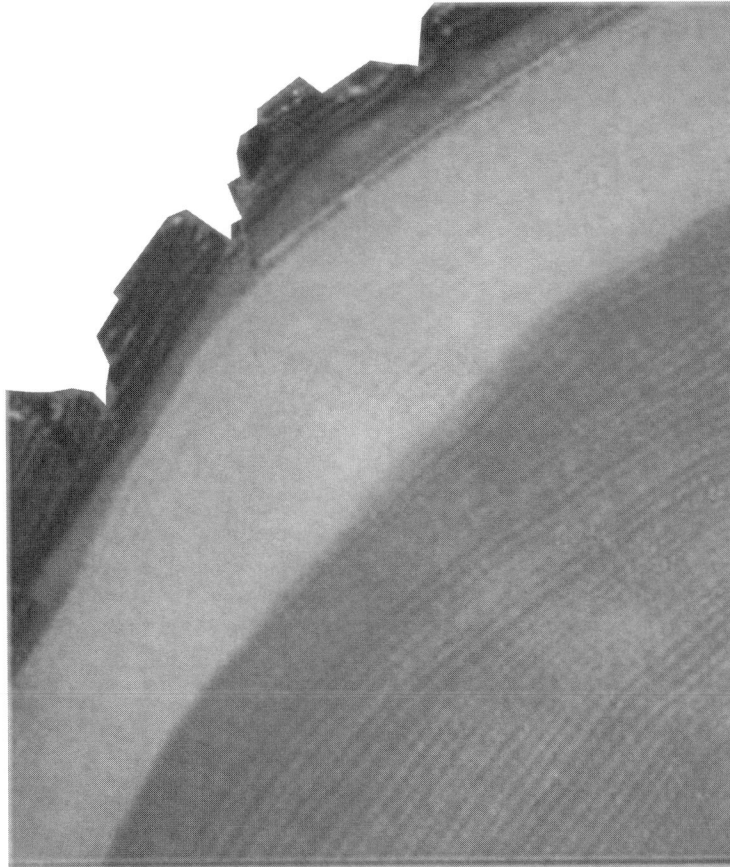

Figure 3.2. Clear distinction between heartwood and sapwood (courtesy TRADA Technology)

Heartwood is metabolically inactive. As the tree ages, it increases in girth as a consequence of the cambium laying down a new layer of sapwood each year. The central part of the trunk begins to lose water and stored food substances. The living cells in this region undergo a slow process of conversion as they senesce and may convert food and waste materials into extractives before eventually dying. This can result in a number of changes within the developing heartwood. The colour of the timber may darken due to the deposition of extractives (discussed in Section 3.1), which vary in composition between species. In addition to the formation of extractives, outgrowths from the cell walls may develop to block the vessels, rendering the heartwood less permeable. These outgrowths are known as tyloses and are a common feature of teak and oak and appear as balloon-like structures under the microscope.

3.3. CLASSIFICATION OF WOOD

The commercial division of timbers into hardwoods and softwoods has evolved from long traditions when the timber trade was dealing with a limited range of species. Today, however, this division bears no relation to the softness or hardness of the timber. The terms 'softwood' and 'hardwood' can be confusing as some softwoods are harder than some hardwoods, e.g. yew, a conifer, is considerably denser than balsa wood, a tropical forest wood. Both groups contain timbers that vary in density, strength and resistance to biological attack, i.e. natural durability.

Furthermore, a single species of timber may grow in various parts of the world where each country may use its own indigenous name. British Standard BS 7359:1991 *Commercial timbers, including sources of supply* gives recommended standard names for most timbers used in the UK, although the specifier should be aware that this standard is not exhaustive.

The differences between softwoods and hardwoods are briefly explained and illustrated in Boxes 3.1 and 3.2. Diagnostic features to identify timber species, such as the orientation of the pits, are also discussed.

Box 3.1. Softwoods

Softwood timber is produced from gymnosperms, the coniferous or cone-bearing trees, which are mostly evergreen. The quality of softwoods depends largely on the proportion of thin to thick walled tracheids and on the contrast between the wood of the early and latewood zones. Distribution of these zones is affected by the duration of the growing season and northern latitudes. For example, softwood originating from Northern Russia, e.g. European redwood, will be characterised by having narrow growth rings and a high proportion of latewood within each growth ring. Softwood, such as Corsican pine originating from the Mediterranean, will have wider growth rings and comparatively less latewood. Generally, in terms of strength and overall quality, softwood originating from northern latitudes is seen as superior to that originating from South West Europe.

One characteristic of many, although not all, softwoods is their ability to produce resin. The resin is formed in parenchyma cells and in some species is stored and transported in resin canals. These canals are not cells but cavities in the wood lined with a sheath of parenchyma cells. These canals are present both horizontally and vertically and often provide the anatomist with a useful identification feature. In simple terms, the resin canals provide a means of response to wounding or mechanical damage by compartmentalising the affected timber in resin and isolating it from surrounding, healthy tissue. In softwoods only two cell types are present:

Box 3.1. Softwoods (continued)

Tracheids

The 'woody' tissue is made up of cells known as tracheids, which are arranged vertically and comprise 95% of the wood volume. These cells are hollow, needle-shaped and generally 2.5 mm–5 mm in length. The length to width ratio is in the order of 100:1. The tracheids are packed close together and resemble a honeycomb when viewed in cross section. Liquids pass from one tracheid to another through microscopic openings known as bordered pits. The configuration and distribution of these pits affects the permeability of the timber which in turn affects the ease with which it can be treated. The earlywood consists of comparatively thin walled, paler tracheids whose primary function is transportation of sap. The latewood tracheids are considerably thicker walled and therefore darker. The function of the latewood tracheids is primarily support.

Parenchyma

The 'non-woody' tissue is known as parenchyma. Parenchyma may be present both horizontally (rays) or vertically (axial parenchyma). These cells are soft and thin walled. The rays form narrow bands of cells radiating outwards from the pith to the cambium and are continuous. Axial parenchyma in softwoods are arranged as isolated vertical series of cells known as strands.

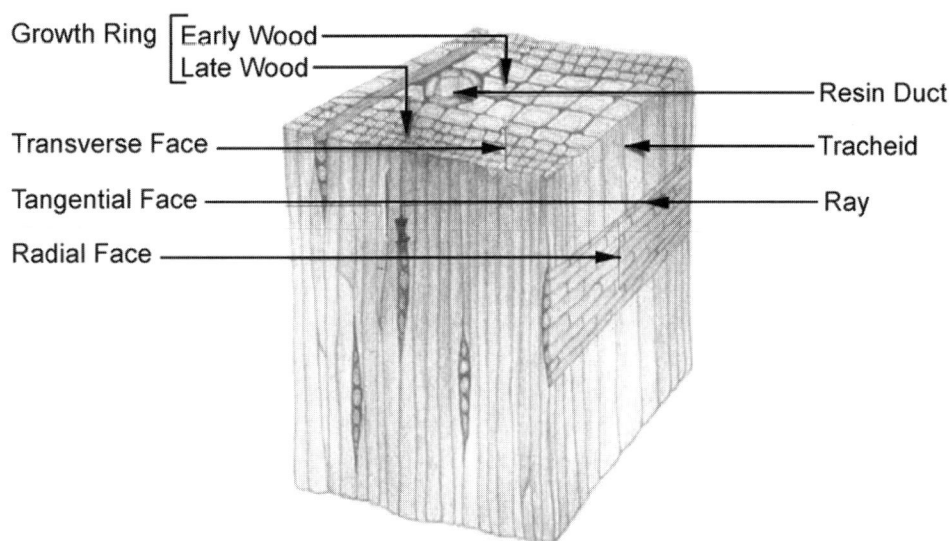

Typical 3-D structure of softwood

Box 3.2. Hardwoods

Hardwood is produced from one group of the angiosperms, known as dicotyledons, some of which are broad-leafed trees. Most tropical hardwoods retain their leaves all year round, while the temperate zone hardwoods are generally deciduous. When first formed from the cambium, the vessel members have end walls just like all other cells, however, early in cell development, the end walls split and are digested by enzymes to form a column of continuous vessels. The split ends of the vessels of different species may vary in their structure. The vessels of some species may be joined through simple perforation plates and in other species only part of the cell wall may have been digested to yield perforation plates. These features often provide the anatomist with a useful diagnostic tool. These plates provide a more effective means of allowing water transport in hardwood than the bordered pit arrangements found in softwood tracheids. Transport between adjacent vessels and ray tissue occurs between numerous pits in the longitudinal walls of vessels. Hardwoods have three cell types:

Fibres

The majority of hardwood tissue is made up of fibres, which have very thick walls offering strength and support to the tree. The fibres are narrow, needle-like cells similar in appearance to the latewood tracheids of the softwoods. The fibre thickness is species dependent and affects wood density.

Vessels

Water-conducting tissue is made up of vessels, which are quite different from the fibres. Vessels tend to be short, perforated elements arranged in axial columns and vary in length from 0.2 mm-0.5 mm and range widely in width from 20 μm-400 μm. The distribution and diameter of the vessels is species dependent and can influence density. Species of timber with many wide diameter vessels, such as obeche, are less dense than species with few, narrow vessels such as greenheart as shown below.

When viewed in the transverse section, these vessels are known as pores. Transport between adjacent vessels and ray tissue occurs between numerous bordered pits in the longitudinal walls of the vessels. For the vast majority of hardwood species there is very little change in the size and distribution of the vessels, except for a reduction in diameter towards the very end of the growth ring. These timbers are known as diffuse porous timbers, examples of which are beech, lime and South American mahogany.

When viewed in the transverse section, some species of hardwood such as oak, elm and ash exhibit two markedly different-sized vessels. Comparatively wider vessels are located in the earlywood band of the growth ring whereas narrower vessels are located in the latewood. Such species of timber are known as ring porous timbers.

In other species such as hickory, walnut and teak the earlywood is marked by incomplete rows of large pores while the latewood appears the same as the ring porous types. These species are classified as semi-ring porous.

Box 3.2. Hardwoods (continued)

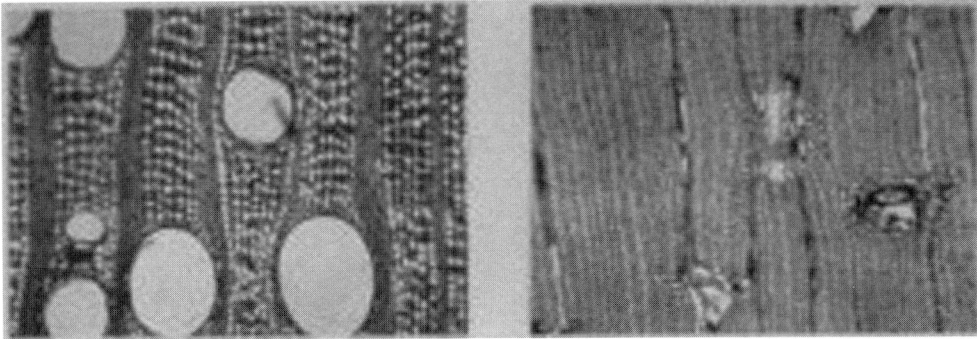

Transverse sections of obeche (left) and greenheart (right) showing differences in diameter of vessels

Parenchyma

In hardwoods the parenchyma tissue is the same as that for softwoods and provides the same function, that of sap storage and conversion. However, the principal difference between the parenchyma of softwoods and hardwoods is that the parenchyma in hardwoods is more abundant and more highly developed, and varies in its distribution and arrangement.

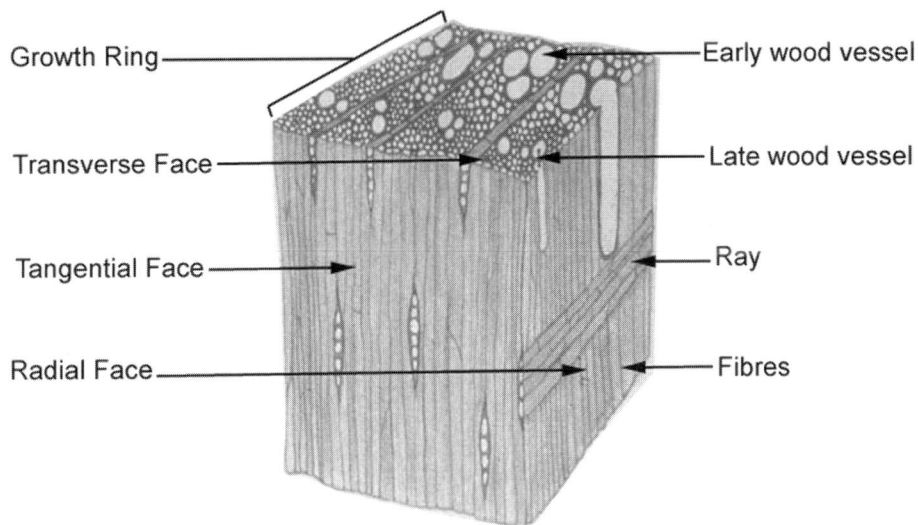

Typical 3-D structure of hardwood

4. *Properties of processed timber*

Most wood is not used in its natural form, but processed to a greater or lesser extent to facilitate its use in structures. The processing activities and their impacts on the properties of timber are described in the following.

4.1. CONVERSION OF TIMBER

Conversion of timber is achieved predominantly by sawing logs into regular sizes and shapes, although some timbers used in heavy civil engineering are also available as hewn baulks. Much of this activity may be undertaken in the producing nation and represents a means of adding value to exported timber as well as making inspection and transportation easier. However, the practices and standards in the producer country may be very different from those common in the country of use, and care must be taken to ensure that designers understand how the timber is produced and converted. For example, greenheart from Guyana is currently produced in imperial dimensions to the Guyana Grading Rules whereas British designers are more familiar with metric dimensions and the grades defined in British Standards.

4.1.1. Timber baulks

Baulks are large section timbers (normally at least 100 × 125 mm) used for piles and other main structural members. The primary requirement for most applications is that the timber should be well grown and straight with the heart of the log surrounded by durable heartwood. This latter feature is often termed 'boxing the heart' and involves producing the baulk in such a way that the pith is present in the centre of the section but not visible on any face or edge (Figure 4.1). This is an important production process because the pith, which is formed in the initial years of the tree's life and makes up the heart of the log, is less dense and less durable. Sections of timber that are to be used in high hazard environments should not display exposed pith and any that do should be rejected or used in less onerous conditions. In addition, the producer should inspect the baulks of timber and reject those baulks with large knots, shakes or other features considered to be unsatisfactory in a structural member.

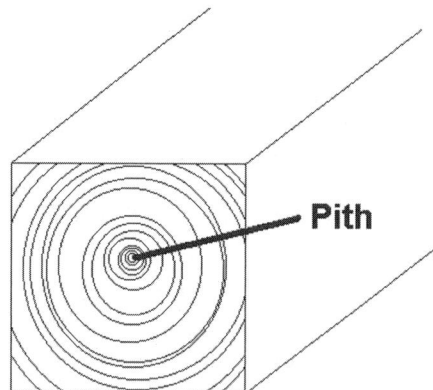

Figure 4.1. Boxed heart: perishable pith has been 'boxed' by durable heartwood (courtesy TRADA Technology)

Hewn baulks

The term 'hewn' normally alludes to greenheart but may be applied to other South American species such as basralocus. Hewn square sections are produced by axing segments into the round log and then hewing parallel to the grain with an adze or broad-bladed axe to produce a uniform section around the heart of the log. All species of trees grow with taper from the butt of the trunk to the crown and although some species have less taper than others (greenheart is renowned for its small taper) none is perfectly cylindrical. Producers recognise that engineers would prefer uniformity in section size along the length of a bulk, but this would involve considerable wastage and may unduly limit the size of timbers available. In order to extract the maximum yield from the forest, hewn baulks that have lost the minimum amount of material may be produced and some taper is usually considered acceptable (in the Guyana Grading Rules, classification GR 02 which relates to hewn greenheart square sections allows a taper of 25 mm in 6.0 m or 1 in 240 and cross-sectional dimensional tolerance of ± 25 mm).

In attempting to limit the taper of the baulk the producer may allow some sapwood and/or wane to be present in the hewn square section, particularly in long lengths required for piling. Despite sapwood being perishable, this does not usually pose a significant problem as the proportion of sapwood is small in comparison to the overall section of the hewn square. A degree of wane can also be tolerated providing it makes up a small proportion of the cross-sectional area of the hewn square. Rejection of any wane and/or sapwood from hewn squares of timber may also render the production of long, straight lengths uneconomic. Although sapwood and/or wane may be present, the producer always endeavours to ensure uniformity between adjacent faces.

In the majority of instances where long piles are required, engineers recognise that operational bending stresses at the tip are not as great as those generated at the butt of the pile. Therefore, in some instances, producers of greenheart piles allow the hewn square to taper off into the round at the tip; in this instance the engineer can ensure design requirements are met by specifying a tip diameter which gives the minimum area necessary.

Sawn baulks

Baulks can also be produced by sawing and whilst this generally produces regular sections with parallel sides it may result in additional waste or inclusion of sapwood when compared with hewn timber. It does, however, result in timbers that have a significantly tighter cross-sectional dimensional tolerance (e.g. sawn greenheart baulks have a tolerance of ±6 mm when produced to Guyana Grading Rules classification GR 01). The process is more mechanised and less labour intensive and can make handling and construction easier. The processing of a single log by sawing can result in the production of a variety of baulk (including one boxed heart) and plank sizes.

4.1.2. Sawn planks

Logs may be converted to planks in a variety of patterns, and the decisions as to the sizes which may be cut and the positions at which the cuts should be made have a great influence upon the efficiency of a sawmill. Timber can be sawn in two distinct patterns, plain-sawn timber and quarter-sawn timber, but for the great majority of engineering applications plain-sawn timber is used.

Plain-sawn timber is defined as timber converted so that the growth rings meet the face in any part at an angle of less than 45°. It is produced by sawing the log through-and-through. In this method, a series of parallel cuts are taken in the general direction of the grain. Conversely, quarter-sawn timber is converted so that the growth rings meet the face at an angle of not less than 45°. Figure 4.2 shows the different methods of converting timber.

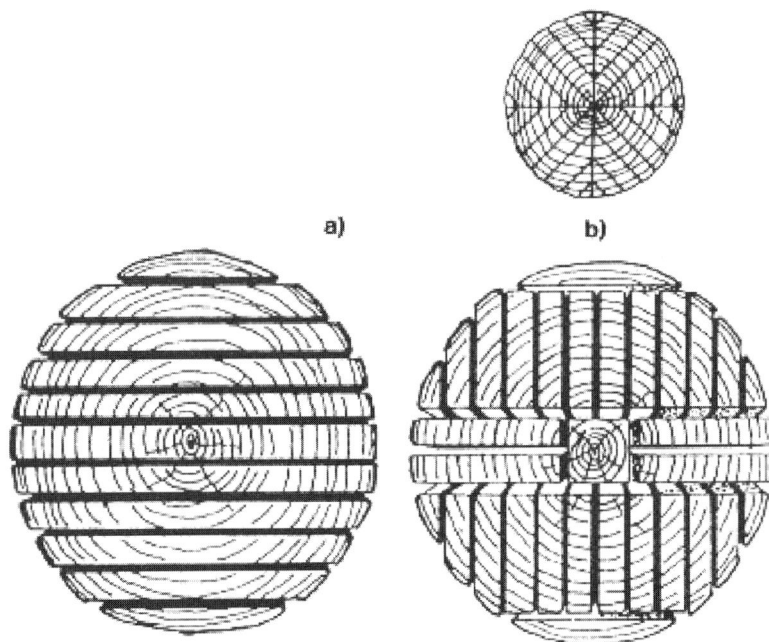

Figure 4.2. Log conversion: a) plain sawn: b) quarter-sawn log (courtesy TRADA Technology)

To produce a significant proportion of quarter-sawn timber, the log must be quartered, that is to say sawn into four parts along two diameters roughly at right angles to one another. Quarter-sawn boards have a number of advantages, including greater dimensional stability and, in tropical hardwoods with interlocked grain, less risk of weakness in tension. However, quarter sawing incurs considerable penalties in the cost of conversion and greater waste. Consequently, nowadays, it is a practice virtually confined to specialist timbers cut for fine craftwork.

4.1.3. Commercial availability

After conversion, hardwoods are frequently traded, both nationally and internationally, as a commodity which is intended to be further processed in a substantial manner. This is in contrast to structural softwoods which, broadly speaking, tend to be less valuable, to come from smaller logs, and to be converted directly into their final cross sections. It is important to appreciate this distinction, since the specifier of constructional hardwoods is more likely to encounter a greater variety of forms of converted timber than when dealing with softwoods.

The term 'dimension stock' is widely applied to such hardwood and although timber can be produced to almost any size required (restricted only by the size of the logs), commercial practice is not quite so random as this implies. For many structural timbers, dimension stock should be readily obtainable in a range of customary sizes (although others may still be available to special order). This is certainly the case for the hardwoods listed with their structural properties in BS 5268: Part 2: 2002 *Structural use of timber, Code of practice for permissible stress design, materials and workmanship*. Customary dimensions and permissible deviations for structural timbers are discussed in Table 4.1. Preferred target cross-sectional sizes for both softwoods and hardwoods are detailed in BS EN 1313: Parts 1 and 2.

Designer expectations may be set by the desire for material uniformity rather than operational necessity, but by gaining an understanding of the methods of production of hewn baulks, the designer may be in an position to relax certain requirements in their specification, provided they do not significantly affect the operational requirements of the design. Normal tolerances for both sawn softwood and hardwood are detailed in BS 5450: 1999 *Round and sawn timber. Permitted deviations and preferred sizes – hardwood sawn timber*, BS EN 1313: Part 1: 1997 *Round and sawn timber. Permitted deviations and preferred sizes – softwood sawn lumber* and BS EN 1313: Part 2: 1999 *Round and sawn timber. Permitted deviations and preferred sizes – hardwood sawn lumber*. The permissible deviations on sawn timber are summarised in Table 4.1.

Table 4.1. Tolerances for sawn hard and softwoods

Section	Hardwoods		Softwoods	
	Dimensions (mm)	Tolerance (mm)	Dimensions (mm)	Tolerance (mm)
Thickness	<32	–1 / +3	<100	–1 / +3
	>32	–2 / +4	>100	–2 / +4
Widths	<100	–2 / +6	<100	–1 / +3
	100-200	–3 / +9		
	>200	–4 / +12	>100	–2 / +4
Lengths	–	+3% of target length but not exceeding 90	–	–

The values cited in BS EN 1313: Parts 1 and 2 are applicable within Europe. However, structural timbers are also sawn in producer nations which may employ their own grading rules with different tolerances to those in BS EN 1313. This should not be seen as a barrier to specifying an appropriate timber but the designer will need to understand the differences and assess whether narrow tolerances are actually required. For example, some greenheart baulks are exported as hewn squares which are often used as marine pilings. Tolerances of hewn timber are not as accurate as sawn timber, but provided the designer considers these greater tolerances, they should not pose a problem for the design of the structure. Hewn faces are usually of sufficient accuracy to permit the use of connectors although special sawing may be undertaken to provide a more even surface onto which other horizontal and diagonal bracing components are fixed. Having considered some of the characteristics that may be evident on long, hewn timber piles, the piles could be part hewn and part sawn as required by top appearance and for fixing horizontal bracing. The piles could also be part sawn with an allowance for wane and part round at the tip providing maximum pile length with least conversion wastage.

Stocks of marine timbers are held by timber importers but it is not unusual to import project-specific requirements as direct importation allows for conversion with minimum wastage and therefore reduces costs. However, as discussed later in Chapter 5, direct importation usually involves a lengthy delivery period (of the order of 3-6 months) and it may be advantageous for the client organisation to pre-order the timber before the construction contract is let.

4.2.   TIMBER SPECIES PROPERTIES

Traditionally the commercial requirements for large volumes, continuity of supply and price coupled with technical requirements for large section sizes, long lengths, high strength and good durability have meant that only a small number of timber species have been used in coastal and river works. These have tended to be tropical hardwoods with a proven track record and, whilst temperate hardwood species such as oak have also been used in some locations, they are increasingly difficult to obtain in large sizes and are not as resistant to marine borers.

Traditional practice for UK marine structures (BS 6349: Part 1: 2000) has depended on using species such as greenheart, balau, oak, Douglas fir, jarrah, opepe or pitch pine for groyne piling. This reliance on a limited number of species can lead to the market exclusion of other timber species with the result that the full value of the exporting nations' forest products industry will not be realised and the diversity of forests damaged. In an effort to address this, lists of species are provided below to allow the identification of a range of appropriate species, with tables of relevant properties (where available) in Appendix 1.

4.2.1. Species traditionally used in the marine and river environments

These lists are by no means exhaustive and other timbers may be suitable. The information presented has been derived from a number of sources including timber importers and literature, much of which was collated in the 1960s and 1970s. The forest industry sector has changed significantly since this time and will undoubtedly change in the future, as different species become available or established species fall from favour. Timber species used in the marine environment are frequently dense, heavy and difficult to work and because of this they are often not viewed as a valuable resource within the producing nation. There is a natural reluctance to specify unknown and untried species and, although it is the great variety of timbers available from tropical forests that makes them so valuable for industrial applications, it also presents an enormous challenge in their use. Marketing lesser known species and encouraging the designer to use an unfamiliar species has always tested the timber industry.

The species lists below include timbers traditionally used in coastal and fluvial engineering as well as lesser known species which have been selected on the basis of expected durability and environmental status. It should be noted that out of approximately 80 species, only 14 tropical hardwoods are identified as being readily available within the UK and these are presented in Table 4.2.

Table 4.2.   Readily available species

Trade name	Botanical name
Tropical hardwoods	
Balau	*Shorea* spp.
Basralocus	*Dicorynia guianensis*
Ekki	*Lophira alata*
Greenheart	*Ocotea rodiaei*
Jarrah	*Eucalyptus marginata*
Karri	*Eucalyptus diversicolor*
Kempas	*Koompassia malaccensis*
Keruing	*Dipterocarpus* spp.
Massaranduba	*Manilkara huberi*
Mora	*Mora excelsa*
Okan (Denya)	*Cyclodiscus gabunensis*
Opepe	*Nauclea diderichii*
Purpleheart	*Peltogyne* spp.
Teak	*Tectona grandis*
Temperate timber spp.	
American/Caribbean pitch pine	*Pinus caribbea/elliotis/palustris*
Douglas fir	*Pseudostuga menziesii*
European larch	*Larix* spp.
European redwood	*Pinus sylvestris*
Oak	*Quercus* spp.

The commercial exploitation of such a narrow range of species can accelerate the depletion of, and escalate the price of, those timber species extracted from tropical forests. In addition, recent moves by the European Union and American authorities to restrict the use of proven CCA (copper/chromium/arsenic) wood preservatives in the marine environment may place further pressure on the species traditionally used. Other species that may be suitable are listed in Table 4.3.

Table 4.3.   Other potentially suitable species

Trade name	Botanical name
Acaria quara	*Minquartia guianensis*
Amiemfo samina	*Albizia ferruginea*
Andira	*Andira* spp.
Angelim vermelho	*Hymenolobium* spp.
Ayan	*Distemonanthus benthamianus*
Belian	*Eusideroxylon zwageri*
Bruguiera	*Bruguirea gymnorhiza*
Brush box	*Tristania conferta*
Chengal	*Balanocarpus heimii*
Courbaril (jatoba)	*Hymenaea courbaril*
Cumaru	*Dipteryx* spp.
Cupiuba	*Tabebuia* spp.
Dahoma	*Piptadeniastrum africanum*
Danta	*Nesogordonia papaverifera*
Denya	*Cylicodiscus gabunensis*
Freijo	*Cordia goeldiana*
Guariuba	*Claricia racemosa*
Heririera	*Heritiera littoralis*
Ipe	*Tabebuia* spp.
Itauba	*Mexilaurus* spp.
Kapur	*Drylobalanops* spp.
Kauta/Kautaballi/Marish	*Licania* spp.
Keledang	*Artocarpus* spp.
Keranji	*Dialum* spp.
Kopie	*Goupia glabra*
Louro gamela	*Nectandra rubra*
Manbarklak	*Eschweilera* spp.
Mangrove cedar	*Xylocarpus* spp.
Muiracatiara	*Astronium lecontei*
Padauk (Andaman)	*Pterocarpus dalgerbioidies*
Piquia	*Caryocar villosum*
Pyinkado	*Xylia xylocarpa*
Red Angelim	*Dinizia excelsa*
Red Peroba	*Aspidosperma polyneuron*
Resak	*Cotylelobium* spp. and *Vaticum* spp.
Rhizophora	*Rhizophra* spp.

Other hardwood species may be suitable for use in areas where there is low/no risk of marine borers, such as fresh water. This list is not exhaustive and there may be, quite literally, hundreds of species with the suitable physical properties for use in river and canal environments. The species given in Table 4.4 are a reflection of a number of species that have been reported to have been used in heavy construction where there is no risk of marine borer activity.

Table 4.4.   Species suitable where there is no risk of marine borers

Trade name	Botanical name
Abiurana	*Pouteria guianensis*
Afzelia	*Afzelia* spp.
Bompagya	*Mammea africana*
Castanha de arana	*Joannesia heveoides*
Determa/red louro	*Ocotea rubra*
Essia	*Comretodendron africanum*
Favinha	*Enterolobium schomurgkii*
Jarana	*Lecythis* spp.
Louro itauba	*Mezilaurus itauba*
Louro preto	*Ocotea fagantissima*
Piquia marfim	*Aspidosperma desmanthum*
Sapupira	*Bowdichia nitida*
Tatajuba	*Bagassa guianensis*
Tetekon	*Berlinia* spp.
Utile	*Entandrophragma utile*
Wallaba	*Eperua* spp.

4.3. MOISTURE IN TIMBER

Water is an essential constituent of timber. Living trees and freshly felled sawn timber can contain large proportions of water – up to about 200% moisture content of the timber. The moisture content figure can be more than 100% because the weight of water in timber is expressed as a percentage of the oven dry weight of the wood, which is determined by using the following formula:

$$Moisture\ Content\ (MC)\% = \frac{weight\ of\ wet\ wood - weight\ of\ dry\ wood}{weight\ of\ drywood} \times 100$$

Thus a piece of wet timber whose weight is half dry wood and half water will have a moisture content of 100%. Above the 25-30% moisture content level, water fills or partially fills the wood cell cavities. When wood dries, this water is lost first. This reduces the weight of the piece but does not change its dimensions. When the cell cavities are empty but the cell walls still retain their bound water, the wood is said to be at fibre saturation point, above which the timber may be described as 'green'. Further drying below fibre saturation point results in shrinkage of the wood as the walls of the woody tissue contract. When dried wood is put into a wet or moist environment this process is reversed. It is usually necessary to dry wood before it is used or processed in some way, i.e. impregnated with wood preservatives under pressure, unless it is going to be used in water or in a very wet environment or section sizes render drying uneconomic.

32

Box 4.1. Timber drying process

The figure below illustrates the process whereby water (marked in black) is lost from drying timber. (A) Shows timber saturated with water, as the timber dries, free water is lost from the lumen of the cells (B) until the timber reaches the fibre saturation point (C). This is the point where there is no free water left in the timber and any further loss of water will now come from the cell wall and will result in shrinkage as illustrated in (D).

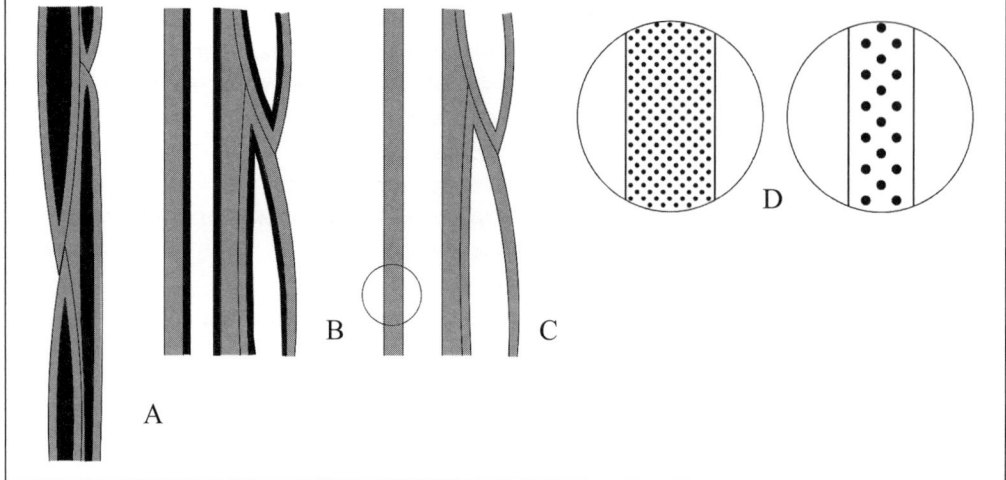

There are two main reasons for drying timber (Box 4.1), which are:

- The timber of many species will decay if kept at high moisture contents for long periods. Some susceptible timbers will suffer from mould discoloration or staining even if they are kept wet for only a short time. Timber below about 20% moisture content is too dry to suffer such discoloration or decay.

- Wet timber will usually dry in service. As it loses water below the 25-30% fibre saturation point it will shrink laterally. If the grain of the piece is not absolutely straight, distortion may occur. Where appropriate, pre-drying the wood allows these inevitable dimensional changes to be avoided in service and enables the production of accurately shaped and sized components.

Further reasons for drying that may be important in specific cases are:

- to save weight during transportation
- to facilitate machining
- to enable strong glue joints to be made
- to allow preservatives to penetrate
- to increase the loads that timber can carry.

With reference to using large sections of naturally durable timber such as ekki or greenheart as piles, planking, wharfing and decking, the timber will almost certainly be in the 'green' state. Components that are introduced to a permanently wet environment, for example tidal and intertidal zones, will maintain a high moisture content above the fibre saturation point, and are unlikely to experience movement in service as a consequence of changing moisture content.

4.3.1. Movement of timber

Movement, i.e. shrinkage or swelling occurs in response to changing environmental conditions such as changes in temperature and relative humidity. The response of timber to changes in temperature and humidity is quite slow and tends to 'average out' minor changes in conditions such as diurnal fluctuations. The outer layers of the timber respond more rapidly to changes than the inner sections of a piece. Protective or decorative coatings, such as paints, varnishes and exterior wood finishes slow down the response but will not prevent the moisture content of the timber from changing. In general terms, changes in moisture content of timber, provided it is permanently out of contact with the ground or water, are measurable on a seasonal basis, rather than in terms of days or weeks.

Actual movement values vary between species and can be influenced by mechanical restraint. For most practical purposes the following assumptions should suffice:

- timber does not shrink or swell lengthwise along the grain
- shrinkage starts as the timber dries below about 30% moisture content or the fibre saturation point.

If timber is put into service at a moisture content higher than that which it is likely to reach with time, for example 'green' ekki pier decking, shrinkage and distortion can occur (Box 4.2).

Box 4.2. Example of timber shrinkage

If an ekki decking board at 45% moisture content is put into an environment where the expected service moisture content will vary between 12-16% it can be expected to shrink by 1% of the board's dimensions across the grain for every 3% drop in moisture content below the fibre saturation point of 30%. Timber will only experience shrinkage once the moisture content falls below the fibre saturation point. Therefore a drop of 18% moisture content may lead to shrinkage across the grain by up to 6% of the dimension of the decking board, i.e. green boards of ekki of width 300 mm could shrink to a width of 282 mm.

Whilst it is unrealistic to apply high precision to matters involving moisture content, a severe mismatch between the moisture content at the time of supply, storage or installation and the timber's eventual service moisture content will often lead to problems in service.

Distortion is caused by the difference in shrinkage in the tangential direction compared with that radially, coupled with the fact that the grain of a piece of timber rarely runs true. Thus, large changes in moisture content below fibre saturation point can result in the bowing or twisting of timber components. However, this can be reduced by specifying and designing for construction with robust fixings at the appropriate frequency to provide restraint. Careful design to accommodate anticipated movement, coupled with sensible moisture content specification will avoid such problems.

A stable moisture content is rarely reached in practice. The change in dimension exhibited by timber after its initial drying shrinkage is termed 'movement'. This varies between species. Table 4.5 indicates the movement values of a number of common species, based on a classification system devised by the Forest Products Research Laboratory. The classes are based on the sum of the tangential and radial movements corresponding to a change in humidity conditions from 90% to 60% relative humidity at a constant temperature which are detailed below.

Small:	1% change in dimension requires 5% change in moisture content
Medium:	1% change in dimension requires 4% change in moisture content
Large:	1% change in dimension requires 3% change in moisture content

Where movement tolerances are critical, a timber with small movement characteristics should be considered.

Table 4.5.   Movement values for different timber species

Movement value		
Small	Medium	Large
Douglas fir, Asian teak, western hemlock, makore, American mahogany, purpleheart, padauk, western red cedar, balau	European redwood, oak, European elm, radiata pine, European larch, sapele, utile, jarrah, greenheart	Ekki, karri, sweet chestnut

4.4.   STRENGTH

The strength of timber may be determined by two different approaches, one of which is based on testing small clear specimens, while the other employs testing actual structural size timbers. The method for testing small, clear specimens is described in BS 373: 1957 *Methods of testing small clear specimens of timber*. This method has been used for many years and forms the basis for many of the published stresses (including, for example, those within the *Handbook of hardwoods*, HMSO, 1997) given for many timber species. This method still remains valid for the strict academic comparison of timber from different trees or different species. Testing small clear specimens requires the preparation and testing of samples of the maximum quality that can be obtained. Therefore these tests are not representative of timber actually used in service. The application of a number of factors that take into consideration the presence of knots and other strength-reducing features permits the derivation of permissible (working) stresses for the species tested as small, clear specimens. The precise method of deriving permissible stresses from ultimate stresses is beyond the scope of this manual but a method of estimating them is described in Box 4.3.

The method for testing actual structural size timbers is described in BS EN 408: 1995 *Timber structures. Determination of some physical and mechanical properties.* A comparison of test data for structural size specimens to small specimens will reveal substantial strength reductions arising from features such as knots and distorted grain. Extensive testing of structural sizes of commonly used softwoods such as European redwood and Douglas fir, for example, have been carried out. The data derived from testing actual structural size components of these, and other softwood species, has resulted in these softwoods being categorised into their respective strength classes. This enables the designer to apply an 'off the shelf' permissible stress for that particular species of softwood. These strength classes and their permissible stresses are presented in BS 5268: Part 2: 2002. However, only a small number of hardwoods have been tested in this manner.

4.4.1. Strength-grading

Since timber is a natural material, some form of selection or grading is necessary to ensure that the timber is used to best effect. Performance is related to choice of grade, as well as the selection of timber species. Over-design and over-specification are wasteful and expensive, both in terms of direct cost and in the overall use of timber resources. Most of the timber-producing areas of the world have some form of quality grading system in operation. These systems not only ensure that the producers obtain a higher price for high quality timber, but also ensure that unnecessary high qualities are not used where they are not required. Grading can be based on appearance of the timber where aesthetic qualities are important or on physical features that affect its strength.

Strength-grading, also known as stress-grading, provides a means of assessing the strength of a piece of timber. Timber may be graded mechanically with strength-grading machines or visually by trained visual strength graders. The strength of timber depends upon many factors, which include growth characteristics such as knots, fissures and slope of grain.

Visual strength-grading rules define the size, type and number of strength reducing characteristics permitted in each defined grade. The grader assesses each piece of timber and stamps those that satisfy the grade requirements with an appropriate mark. Both hardwoods and softwoods may be visually graded.

Machine strength-grading usually only applies to softwoods. The strength grade is based upon the relation between strength and stiffness. Each piece is mechanically tested and visually assessed. Pieces that comply with the grading rules are marked with an appropriate stamp.

The strength of timber is species related. Therefore it is important to identify the species of timber that is being graded. This is because different species may ultimately be assigned to different strength classes. In other words, there is a wide range of combinations of species and grade. To simplify design, species/grade combinations of similar strength are grouped together into strength classes which are defined in BS EN 338: 1995 *Structural timber. Strength classes.* Guidance for the use of timber in construction in the UK may be taken from the following standards.

- BS 5268: Part 2: 2002 *The structural use of timber. Code of practice for permissible stress design, materials and workmanship*
- DD ENV Eurocode 5: 1995: Part 1 *Design of timber structures. General rules and rules for buildings.*

Strength-grading softwood

The strength classes under which softwoods are classified are: C14, C16, C18, C22, C24, C27, C35 and C40. C40 is the highest strength class for structural softwood. Grades of softwood above C30 are unusual. In commercial terms, C16, C24 and C27 are the most common grades available. The 'C' stands for coniferous. With reference to visual grading of softwood, there are two visual strength grades, which are:

- GS – General Structural grade
- SS – Special Structural grade.

The relevant standards referring to softwood strength-grading are:

- BS 4978: 1996 *Specification for visual strength grading of softwood*
- BS EN 519: 1995 *Structural timber. Grading requirements for machine graded timber and grading machines.*

Strength-grading hardwood

In general terms, hardwoods in the medium to high-density range (>550 kg/m^3) are superior in terms of strength and stiffness to softwoods, and are usually available in longer, larger sections.

Hardwoods may be visually graded using the guidelines given in BS 5756: 1997 *Specification for visual strength grading of hardwood.* There are five visual strength grades, which are:

- HS – Tropical hardwoods
- TH1 and TH2 – General structural temperate hardwood for timber of a cross-sectional area of less than 20 000 mm^2 and a thickness of less than 100 mm. TH1 is a superior grade to TH2
- THA and THB – Heavy structural temperate hardwood for timber with a cross-sectional area of 20 000 mm^2 or more and a thickness of 100 mm or more. THA is a superior grade to THB.

The strength classes and their respective permissible stresses under which hardwoods are classified in BS 5268: Part 2: 2002 are D30, D35, D40, D50, D60 and D70. D70 is the highest strength class for structural hardwood. The 'D' stands for deciduous. Timber species that meet the visual strength grading requirements detailed in BS 5756 may be categorised into their respective strength class by species.

Tropical hardwoods may also be graded in producing countries using local systems, (such as the Guyana Grading Rules) which are not always entirely consistent with UK or European standards. Nevertheless, many engineers have developed experience of the producing nation rules and are prepared to use them. Alternatively, timber may be inspected and/or re-graded on arrival in the UK.

4.4.2. The use of wet or dry grade stresses

Unless timber is in contact with water or exposed to very damp conditions, its moisture content will often stabilise to a level well below the fibre saturation point, typically varying between 10% and 20% in the UK. In some cases the moisture content may rise above 20% if the components are exposed to wetting for prolonged periods.

For service conditions where the timber is in a wet environment, the 'wet' stresses should be used. This is because 'wet' timber is weaker than dry timber and most published grade stresses refer to dry timber. 'Wet' stresses may be obtained by applying modification factors, or K_2 factors to the quoted dry stresses. Table 4.6 summarises the K_2 factors, which are detailed in BS 5268.

Table 4.6.   Modification factor K_2

Property	Value of K_2
Bending parallel to the grain	0.8
Tension parallel to the grain	0.8
Compression parallel to the grain	0.6
Compression perpendicular to the grain	0.6
Shear parallel to the grain	0.9
Mean and minimum modulus of elasticity	0.8

It is advisable to use these K_2 factors when designing with timber sections in excess of 100 mm in any dimension, irrespective of exposure conditions, unless special provision has been made to source dry timber. It is uncommon to obtain dry timber that is thicker than 100 mm because it is not economic to attempt to kiln-dry timber of this thickness. It may take four years to air-dry timber of 100 mm thickness. When air-drying timber, the rule of thumb is that it takes one year per 25 mm (one inch) of thickness to dry.

Box 4.3.   Estimating the permissible stresses for timber species

'Off the shelf' design values or permissible stresses categorised into strength classes may be obtained from BS 5268: Part 2: 2002 (factors determining permissible stresses are described in greater detail in Section 6.6.2). Species of timber are categorised into their respective strength classes which take into account the variation in the type and quality of the timber available. Strength classes group together grades and species with similar strength properties.

The strength class system enables the designer to specify a certain strength class without having to choose a specific species. The advantages of the strength class

---

Box 4.3.   Estimating the permissible stresses for timber species (continued)

system is that additional species, fit for purpose, within the same strength class can be incorporated into a scheme without affecting the existing design. At the time of design, the engineer may not need to be aware of costs, availability and sizes of a specific species and by choosing a strength class, the tenders can be used to select the most suitable and economic species or grade of species on offer.

The strength class system is described in BS EN 338: 1995 *Structural timber. Strength classes* and is based upon the strength, stiffness and density value for each class. However, BS 5268: Part 2: 2002 categorises only a limited number of species of hardwoods into their respective strength classes which are summarised below.

Species of hardwood cited in BS 5268: Part 2: 2002 and categorised by strength class when graded in accordance with BS 5756: 1997

Species	Average density kg/m^3	Strength class
Balau	825	D70
Greenheart	1030	
Ekki	1051	D60
Kapur	760	
Kempas	850	
Karri	880	D50
Keruing	740	
Merbau	800	
Opepe	740	
Iroko	640	D40
Jarrah	880	
Oak	720	
Teak	650	
Oak	720	D30

Although the above list is limited, it does show a range of important medium-high density hardwoods which have been, and are imported into the UK and used for building purposes. Furthermore, heavy constructional timbers such as balau and greenheart have also been included in this table. Oak may be assigned one of two strength classes and this is dependent upon the timber meeting the grading requirements for the appropriate strength class. Tropical timbers are graded on a pass/fail basis and may only be assigned the strength class appropriate to that species.

There are many well known and lesser known hardwood species that may be suitable for marine and coastal engineering but they have not been categorised into their respective strength classes, i.e. no 'off the shelf' permissible stress values are available. The designer may have only the ultimate stresses available as a design

Box 4.3. Estimating the permissible stresses for timber species (continued)

guide. As stated earlier, ultimate stresses are values derived from testing small, clear and defect-free sections of the candidate species and are higher than the strength class values found in BS 5268: Part 2: 2002 which take into account the variation and occurrence of strength-reducing features in the timber.

Assuming that the designer is able to obtain ultimate stress values for the candidate species and is reasonably confident of the pedigree of these data, the appropriate strength class for a hardwood species not cited in BS 5268: 2002 may be estimated using the following method although it must be borne in mind that this method is not a precise engineering tool.

The first step is to match the density of the lesser known species against that of a well known species, ideally one that is cited in BS 5268: 2002. Once this has been undertaken the designer can then derive a conversion factor for a key physical attribute of the established species by using the following formula:

(permissible stress value, by strength class, for bending strength / ultimate stress value for bending strength) = k

This conversion factor k can then be applied to the ultimate stress data for the lesser known species (or species not cited in BS 5268) and thus derive an estimated permissible stress. The designer should be aware that the relationship of one strength property to another is not completely constant from one species to another. Cases may arise where a timber would be classified as belonging to one strength class according to one property, but a different class according to another. These cases may be overcome by considering key attributes as more important than others. For example, bending strength may be considered as being an essential characteristic to be satisfied for a particular strength class whereas shear strength may only have to approach within a certain percentage of the strength class boundary deemed by the designer to be acceptable.

The next step is to ensure that the candidate species can be visually strength graded in accordance with the exporting nation's grading rules or graded in accordance with the requirements detailed in BS 5756: 1997 *Visual strength grading of hardwood*. This ensures that timbers exhibiting unacceptable strength reducing features such as slope of grain and knots are rejected.

Example: Comparison of greenheart to cumaru

Species	Density kg/m^3	Ultimate bending strength at 12% moisture content N/mm^2	Permissible bending strength at 12% moisture content N/mm^2 (Strength class as per BS 5268)	Conversion factor k	Strength class
Greenheart	1030	181	23	0.127	D70
Cumaru	1200	199	Unknown	0.127	unknown

For the designer to estimate the permissible bending strength and appropriate strength class for cumaru, the conversion factor derived for greenheart (0.127) is applied to the ultimate bending strength for cumaru. This results in an estimated permissible bending strength of 25.2 N/mm^2. In terms of bending strength, cumaru can be assigned the permissible stresses for strength class D70. A similar exercise can be carried out for danta.

Box 4.3. Estimating the permissible stresses for timber species (continued)

Example: Comparison of greenheart to danta

Species	Density kg/m^3	Ultimate bending strength at 12% moisture content N/mm^2	Permissible bending strength at 12% moisture content N/mm^2 (Strength class as per BS 5268)	Conversion factor k	Strength class
Greenheart	1030	181	23	0.127	D70
Danta	1000	137	Unknown	0.127	unknown

For the designer to estimate the permissible bending strength and appropriate strength class for danta the conversion factor derived for greenheart (0.127) is applied to the ultimate bending strength for danta. This results in an estimated permissible bending strength of 17.4 N/mm^2. When compared with the minimum permissible stresses for the strength classes detailed in BS 5268: Part 2: 2002, in terms of bending strength, danta can be assigned the permissible stresses for strength class D50.

Where ultimate stress values are not available for lesser known species or there is little confidence in the reliability of the data, then density may be used to gain a 'ball park' estimation of strength. Density is an important factor that can influence the strength and stiffness of timber and correlates well with certain properties such as bending strength, hardness and compression strength parallel to the grain. However, these estimations should only be used for comparative purposes. The strength of timber is obviously important, forming one of the key design parameters, but since timber is a natural material, there is some variability even within single species.

4.5. DURABILITY

Timber is an inherently durable material, which is resistant to most biological attack *provided it remains dry*. However, prolonged wetting leads to a risk of decay by wood rotting fungi, although susceptibility varies according to the wood species, and this is known as '*natural durability*'. Resistance to mechanical damage also varies according to the species.

4.5.1. Mechanical damage

Timber is a variable material and its natural durability is affected by many factors. Two problems occur when timber is used in marine construction: biological attack and mechanical abrasion. The latter is a significant problem for structures located on mobile beaches – particularly those with a low tidal range since this concentrates the attack on a small part of the structure. Abrasion is caused by impacts from beach material carried around or over the structure, predominantly by waves. It can result in significant loss of section of timber components and significantly reduce the service life of structures.

The abrasion resistance of different species of timber in this type of situation may be influenced by the effect of marine soft rot fungi and also by the extent of saturation of the timber. Timber with a high moisture content (i.e. above the fibre saturation point) is weaker than dry timber. The action of soft rot fungi can cause softening at the surface of the timber, accelerating the rate of mechanical abrasion as the decayed timber is easily eroded.

The effects of abrasion can be limited by the selection of durable timber, but the detailed design of the structure and provision of sacrificial sections to protect exposed structural components can also be very effective. Since the purpose of groynes is to reduce the transport of beach materials, they can be particularly susceptible to abrasion and further details of the problem and counter measures are discussed in Section 6.2.7.

Mechanical abrasion of timber may be uneven and this may be a reflection of beach conditions or variation in the natural properties of the timber species such as density, for example. The greatest abrasion usually occurs in the intertidal zone between the lower limit of shingle and the average high tide level. The rate of wear will vary from beach to beach and from species to species. Once installed, however, it is possible to estimate the rate of wear of the timber components by carrying out annual inspections of the structures and thereafter taking remedial measures with sacrificial pieces to minimise wear problems.

Different species of timber will have varying resistance to abrasion. There is no standard test to determine the abrasion resistance of timber species but comparative tests have, and can be, carried out on sites where abrasion is a recognised problem. In some instances, field testing may not be practicable so the designer will have to rely on species of timber with a proven track record. Reliance on a narrow range of species is a problem that has been discussed earlier (see Section 4.2.1) as potential candidate species which have not been accepted within the industry may be rejected by the designer as there is little knowledge regarding abrasion resistance.

It is possible to make a fairly accurate judgement as to the abrasion resistance of candidate species by assessing alternative mechanical properties. Density and hardness are two properties which can provide an indication as to the abrasion resistance of candidate species. The greater the density and hardness, the greater the probable abrasion resistance. For candidate species of similar density, the species with the greater hardness will probably have better abrasion resistance.

4.5.2. Biological attack

Durability may be a natural feature of the timber or it may be imparted by the use of preservative treatments. However, before discussing the technical factors, critical to the successful preservation of timber, it is important to gain an understanding of the attack mechanism and natural durability.

Fungal decay and insect attack

Wood rotting fungi obtain their food by breaking down wood cell walls. They can cause staining, decay, loss of strength and can cause complete disintegration of the timber component. For fungal decay to occur, the timber must have a moisture content in excess of 20%. The first line in defence in preventing fungal decay is to keep the timber below this decay threshold. This is not possible in timber for use in the marine and fresh water environments. Therefore, reliance must be placed on good design and the specification of naturally durable timber species or timber that has been treated in the appropriate manner with a wood preservative.

There are two types of wood rotting fungi that can affect the strength of timber components in marine and fresh water environments. These are soft rot and wet rot type fungi. The main risk of fungal decay affecting timber structures in the marine and fresh water environments comes from wet rot type fungi attacking timber within and above the intertidal zone. Such fungal attack can cause extensive pockets of decay and severe loss in strength. As wet rot type decay develops, it usually does so at an ever increasing rate with the result that the subsequent loss in strength of the component also occurs at an ever accelerating rate. Soft rot fungi will attack timber that is immersed in water. Soft rotting type fungi erode the outer layers of the timber components at a relatively slow rate. The outer surfaces are typically dark, soft and 'cheesy' in texture. Therefore regular inspections of the structure should be incorporated as part of any maintenance regime.

The presence of insect attack in marine and fresh water constructions is usually limited to the perishable sapwood section of a component or can be symptomatic of fungal decay being present.

Marine borer attack

The principal agents causing biodeterioration of timber in the marine environment are marine borers, which can cause severe damage in relatively short periods of time. In UK and other temperate waters two types of borer are of greatest significance; the mollusc *Teredo* spp. (shipworm) and the crustacean *Limnoria* spp. (gribble).

Descriptions of *Limnoria* spp. and *Teredo* spp. are provided in Box 4.4 and Box 4.5.

Box 4.4. *Limnoria* spp. (gribble)

Crustaceans are very different from molluscan borers, principally in that crustaceans may move to fresh wood whereas molluscs remain in their burrows for life. In UK waters, the principal crustacean marine borer is *Limnoria* spp. otherwise known as the gribble. Attack by the gribble is superficial and results in the creation of a network of galleries varying in 1-3 mm in diameter at or just below the surface of the wood. These tunnels are characterised by regularly spaced respiration holes at the wood surface.

Adult gribble have whitish-grey segmented bodies with seven pairs of legs and vary from 2-4 mm in length. Ecological factors such as sea temperature and salinity can have a significant impact on the distribution and vigour of the gribble. Extensive attack by the gribble weakens the wood to such an extent that the surface layers are eroded away by tidal action and the rate of erosion may be further accelerated by the action of soft rot type fungi. Gribble are more resistant to pollution and changes in salinity than shipworm, but are commonly found in water warmer than 5°C and salinity exceeding 10%.

Gribble (courtesy "Nature")

Box 4.5. *Teredo* spp. (shipworm)

There are two types of molluscan bivalve borer: the teredinids (*Teredinidae*) (otherwise know as the shipworm) and pholads. The teredinids are the largest and widely distributed group whilst pholads are more restricted to warm temperate and tropical waters and are not thought to have caused significant degradation of timber around UK waters.

Teredinids possess a soft worm-like body with two shells at the head of the animal, which allows the animal to bore into the timber element, usually along the grain. The posterior end of the animal remains in contact with seawater via a small hole of 1-2 mm in diameter. Two siphons, scarcely visible to the human eye, protrude from the hole enabling water to be drawn in providing oxygen and micro-organisms on which the animal feeds and releasing waste and reproductive gametes/larvae. The animals are normally confined to marine and estuary areas with good water-quality, temperature (warmer than 5°C) and salinity exceeding 7%. Their occurrence in the UK is thought to be sporadic and in some cases seasonal. The last comprehensive survey determining the occurrence of marine borers around the UK coastline was carried out by TRADA in the 1960s.

Box 4.5. *Teredo* spp. (shipworm) continued

The precise mode of settlement of juveniles on to wood is disputed. One school of thought is that colonisation is associated with softened regions of wood which have suffered fungal and bacterial attack. Another view is that teredinids will colonise timber irrespective of biological pre-conditioning. The grinding action of the shells produces fine fragments of wood which are ingested and the cellulose component digested.

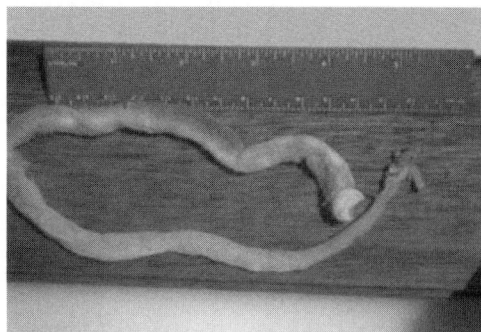

Teredo spp. ('shipworm') *(courtesy Simon Cragg)*

The animal remains in the same tunnel throughout its life and lines it with a calcareous deposit. The animals avoid intruding into neighbouring tunnels and in warmer waters some species can grow up to 2 m in length. Shipworm is extremely difficult to detect and survey methods are detailed in Chapter 9.

Natural durability

Timber species differ markedly in their resistance to biological attack and natural durability may be defined as the natural resistance of the timber to such attack. It is important to recognise that this term only refers to the heartwood of timber species. The sapwood of all timber species has no natural durability. The heartwood of some species, e.g. greenheart, will last for decades in ground and/or marine contact whereas that of European redwood will suffer destruction in a comparatively short space of time. Historically, natural durability was recognised through experience of working with various species. However, in the nineteenth century field tests provided specifiers with more reliable test data enabling categorisation of the timbers' natural durability.

In more recent times natural durability ratings have been determined by one of two methods. Ratings can be based on field tests where 50 mm x 50 mm stakes of heartwood have been exposed in ground contact and assessed at regular intervals. Larger section sizes will last longer in ground contact. For example if a 50 mm × 50 mm heartwood stake of oak lasts 15-25 years in ground contact, it follows that a 100 mm × 100 mm stake can be expected to last 30-50 years. The drawbacks of this test method are that it is time consuming and subject to human error.

A faster, more objective method to assess durability is to carry out weight loss tests with specific laboratory organisms as detailed in BS EN 350: *Durability of wood and wood-based products. Natural durability of solid wood*, Part 2: 1994 *Guide to natural durability and treatability of selected wood species of importance in Europe.* This presents a list of 20 softwoods, 107 hardwoods and 7 commercial species groupings.

BS EN 350 identifies five natural durability ratings. The equivalent British Standard BS 5589: 1989 *Preservation of timber* also classifies timber into five durability ratings. The durability ratings are summarised in Table 4.7. For a wider range of species the reader should refer to BS EN 350: Part 2 and BS 5589.

The natural durability ratings defined in BS 5589 and BS EN 350 do not consider the resistance of the timber to marine borer attack. It is important to recognise that resistance to marine borers does not necessarily parallel resistance to fungal decay. In addition, natural durability in any climatic or geographical region may vary according to the vigour of wood destroying flora and fauna. It is important therefore to carry out a risk assessment based on local information about the performance of wood species and the presence of wood destroying organisms.

Table 4.7.   Classification of natural durability

BS EN 350		BS 5589	Approximate life in ground contact	Examples
Class 1	Very durable	Very durable	>25 years	Jarrah, greenheart, iroko and ekki.
Class 2	Durable	Durable	15-25 years	European oak, sweet chestnut, robinia and yellow balau
Class 3	Moderately durable	Moderately durable	10-15 years	Caribbean pitch pine, Douglas fir, sapele and European larch
Class 4	Slightly durable	Non-durable	5-10 years	European redwood/whitewood
Class 5	Not durable	Perishable	> 5years	Beech, sycamore and ash

Natural durability may also vary within the tree itself, especially for those timbers that are classified as being very durable. In some species such as ekki, some anatomists have described a layer of timber between the sapwood and heartwood as comprising 'transition wood' which is not as durable as fully formed heartwood. BS EN 350 recognises this condition in ekki. With reference to BS EN 350: Part 2 the transition wood in ekki is classified as moderately durable. Transition wood is generally slightly paler than the fully formed heartwood.

As previously stated, it is important to recognise that natural durability against fungi does not necessarily indicate durability against marine borer attack. The majority of timber species are susceptible to marine borers but a limited number of species are recognised as being resistant, i.e. naturally durable in the marine environment. BS EN 350: Part 2 identifies a number of timber species and recent research has investigated the resistance of a number of traditional and lesser known species to attack by marine borers which is described in Box 4.6.

Box 4.6.   Resistance of timbers to marine borers – laboratory trials

Currently, BS EN 275: 1992 *Determination of the protective effectiveness against marine borers* details the protocol for establishing preservative efficacy in the marine environment and may be used to form the framework to assess natural resistance of timber species to marine borers. However, BS EN 275 specifies a five year test period: too long a period for screening tests to be economically viable, especially if producers are encouraging designers to specify lesser known species. In order to address this problem a new test protocol has been developed to prove the potential suitability of candidate species through a rapid laboratory assay. If the candidate species fails in the laboratory test, it will almost certainly fail if 'field' tested.

There are two types of marine borer present in UK waters. The crustacean *Limnoria* spp. (gribble) is ubiquitous whereas the mollusc *Teredo* spp. (shipworm) tends to be limited to the south west coast. In this pilot study, durability was assessed by measuring the rate of digestion of samples by the marine borer *Limnoria quadripunctata* under forced feeding conditions over a number of days. The aims of the pilot study were two-fold: to determine the resistance to *Limnoria* spp. of lesser know species of timber thought to be suitable for use in the marine environment and to investigate the feasibility of introducing a new fast track screening protocol for suitable candidate species.   The preliminary results of this pilot study are summarised below.

Attack on a range of timbers over 15 days compared with rate for Scots pine sapwood

Species	Rank	% Red.	Species	Rank	% Red.	Species	Rank	% Red.
Bruguiera *spp.* (mangrove)	1	100	Uchi torrado	15	74.9	Balau	29	17.2
Cupiuba	2	98.6	Jatoba	16	73.4	Louro petro	30	10.1
Acaria quara	3	95.8	Angelim vermelho	17	69.9	Oak	31	5.4
Wonton	4	94.3	Robinia	18	68.5	Louro itauba	32	4.8
Ekki	5	94.1	Purpleheart I	19	67.9	Scots pine sapwood CONTROL	33	0
Louro gamela	6	91.9	*Rhizophora* spp. (mangrove)	20	65.8	Piquia marfim	34	-4.4
Ayan	7	85.4	Purpleheart II	21	55.9	Jarana	35	-12.7
Denya	8	84.5	*Xylocarpus* spp.	22	47.3	Sapupira	36	-22.2
Piquia	9	83.2	Muiracatiara	23	44.3	Castanha de arara	37	-41.4
Favinha prunhela	10	81.7	Dahoma	24	42.5	Abiurana	38	-41.8
Ipe	11	80.7	Guariuba	25	41.2	Tetekon	39	-42.9
Cumaru	12	80.1	Greenheart	26	41.0	Favinha spp.	40	-74.7
Amiemfo samina	13	80.0	Bompagya	27	30.5	Essia	41	-144.3
Massaranduba	14	76.6	*Heratiera* spp. (mangrove)	28	22.8		42	

% Red. Refers to the % reduction in rate of digestion when the results for candidate species are compared to Scots pine sapwood control blocks. Negative values mean that for that particular timber, animals produced more pellets than for the control indicating that the candidate species was more vulnerable to gribble attack.

Box 4.6. Resistance of timbers to marine borers – laboratory trials (continued)

These initial data have revealed interesting results and have identified a number of potential species that may be used in the marine environment. Several commercially available candidate species from South America have been identified such as cupuiba, louro gamela, cumaru and piquia, for example. In addition, wonton denya (okan) and ayan from West Africa have also been identified as potential species. At the time of writing, further research is being carried out. The initial theory drawn from this research is that if the candidate species is considered to be vulnerable to gribble under laboratory conditions, it is probable that this species will fail against gribble at a faster rate in service than a species with an established track record. However, due to the small sample size and inherent assumptions the results should not be regarded as anything other than indicative. Resistance to *Limnoria* spp. does not necessarily indicate resistance to *Teredo* spp. Alternative test methods to those described in BS EN 275 designed to assess the resistance to *Teredo* spp. are being investigated at the time of writing.

It is also important to recognise that local site conditions will also affect the longevity of timber structures. For example, a timber that is naturally resistant in temperate waters may be more vulnerable in tropical waters. In addition, brackish sites such as estuaries frequently support a greater hazard of shipworm than saline waters. Timber species with a silica content of 0.5% or greater are thought to have resistance against marine borer attack. It is thought that a silica aggregate deposited in certain timber species has a blunting effect on the boring apparatus of marine borers. Resistance may also be imparted by toxic extractives present in the timber (see Table 4.8).

Table 4.8.   Natural defence mechanisms against marine borer attack

Defence mechanism	How	Disadvantages
Silica	Silica aggregate form a mechanical resistance	Difficult processing of timber Silica content has to exceed 0.5%
Alkaloids	Natural toxic extractives	Alkaloids can leach out after several years Eco-toxicity of leaching out of alkaloids is not known
Density and hardness	Great strength and coherence form a mechanical resistance	Difficult processing of timber Performs best in combination with other defence mechanisms

Details of the actual incidence of marine borer attack in the UK and methods of mitigating are given in Section 6.2.2.

4.6. PRESERVATION OF TIMBER

Historically, the wood preservation industry has enjoyed considerable success in prolonging the service of timber with a range of wood preservatives. However, some of these compounds have been banned from use and others are being restricted. The principal reason for the banning and restriction of certain compounds is their reported eco-toxicity. Greater environmental awareness and the need to develop environmentally benign preservatives have driven preservation research and, since the 1970s, many new chemicals and products have been tested for their efficacy. Some commercially available treatments are still being appraised for their long term performance.

The primary objective of the industrial pre-treatment, or pressure treatment of timber with preservatives, is to ensure that even when timber is serving in a hazardous environment it remains sound throughout the design life of the component. Therefore, where naturally durable timber species are not available, it is possible to use non-durable preservative treated timber species in their place. However, the specifier must be aware of a number of limitations governing the use of preservative treated timbers in the marine and fresh water environment, for example, preservative efficacy and timber permeability.

In simple terms, pressure treatment encapsulates the timber component with a preservative envelope which prevents access by fungi or other wood destroying organisms to the vulnerable, untreated core of the component. The treatability of the timber governs the depth of this preservative envelope and the subsequent end use of the component. Pressure treatment is the most effective method of application for the majority of wood preservatives although, by itself, pressure treatment is by no means a universal solution to increasing the longevity of non-durable timber species in hazardous environments.

4.6.1. Types of timber preservative and their treatment methods

Three types of preservative products may be considered appropriate for non-durable timber that is expected to serve in hazard classes HC3, HC4 or HC5, (see Chapter 6 for explanation). These are:

- water-borne preservatives applied by high pressure treatment process
- solvent and water-borne preservatives applied by low pressure treatment process
- tar-oil preservatives.

Once again applied by high pressure treatment process. As well as considering the most suitable formulation, it is also important to specify a timber that is permeable to wood preservatives, especially if the treated timber is to be used in a marine environment, where high preservative loadings are desirable. Such selection must, however, take account of the relevant physical requirements such as strength and resistance to abrasion.

Water-borne preservatives applied by high pressure treatment process

There are two main types of water-borne preservatives applied by high pressure treatment processes. The first type is a traditional preservative based on a chromated/copper/arsenate (CCA) formulation. The high pressure industrial treatment process forces the active ingredients deep into the timber structure where they become chemically 'fixed' to the timber cells and remain leach resistant after treatment. CCA preservatives provide a highly effective protection against all forms of decay and insect attack for timbers used in all hazard classes HC1-HC5. CCA preservatives have been used safely and effectively for over 50 years including the renowned Tanalised® timber, for example. However, since January 2004 the European Commission has restricted the use of preservatives containing CCA. Therefore, preservative manufacturers are now offering new chrome-and-arsenate-free products, which are proving to be effective for applications in hazard classes HC1-HC4. These products, based on copper and a secondary biocide, provide an alternative water-borne preservative for high pressure treatments. Treatment by either type of preservatives results in a pale green/brown colouring to the treated timber.

Typical applications and recommendations for high pressure treatments are detailed in Table 13 of BS 8417 and are a guide to typical schedules. They do not form the basis of a guarantee that specific penetration classes and retentions have been met.

Figure 4.3 illustrates and describes a typical high pressure or 'full cell' treatment process that delivers water-borne preservatives into timber for use in high hazard situations. Where the hazard is not so severe or permeable timber is being treated, a modified treatment process called the 'empty cell' process is used. No vacuum stage is applied at the start of this type of treatment and air in the wood is compressed during the pressure phase. When the treatment vessel is drained of preservative, the air in the timber expands, forcing out more preservative. This results in lower retention values than for the 'full cell' process, but can have significant cost-saving benefits when high loadings of preservative are not required.

® Tanalised is a registered trademark of Arch Timber Protection.

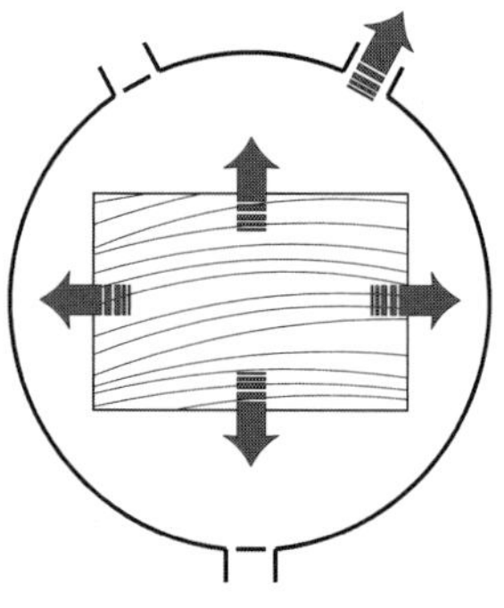	Timber is loaded into the industrial treatment vessel and a vacuum is applied to remove air from the timber. This vacuum period is varied according to the treatability of the timber species.
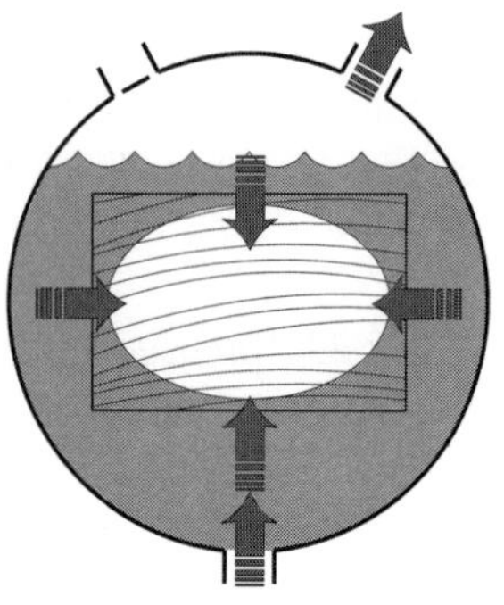	Treatment vessel is flooded with preservative at the required concentration of active ingredient.
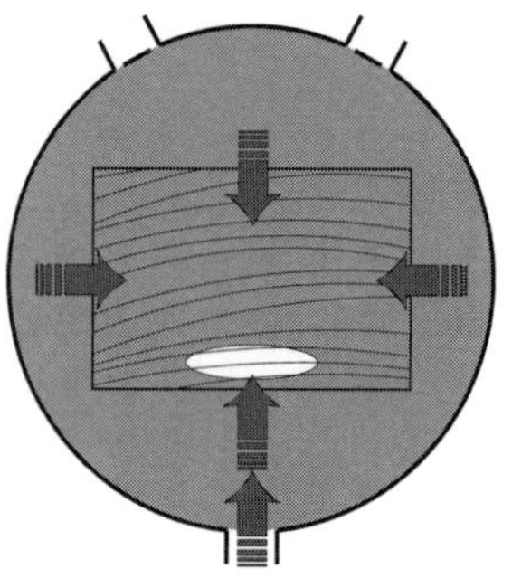	Pressure is applied for the required period so that the necessary penetration and retention values are achieved.
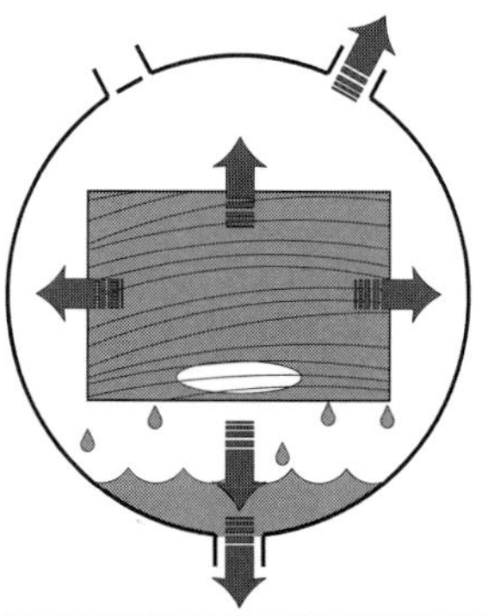	The treatment vessel is drained of preservative and a final vacuum is applied to remove excess preservative from the timber. When vented to the atmosphere, low pressure inside the timber also draws in surface preservative. The timber is left to dry and for the preservative to become fully 'fixed' within the timber cells. Timber treated with CCA type preservatives must be held for at least 48 hours after treatment and until surfaces are dry before the timber can be moved. This allows excess preservative to be collected in an enclosed area.

Figure 4.3. Summary of the 'full cell' high pressure treatment process (diagrams courtesy of Arch Timber Protection)

Light organic solvent and micro-emulsion based preservatives applied by low pressure treatment process

These preservative products can be used to protect timbers used in hazard classes HC1 and HC2 (internal building timbers) and also in hazard class HC3 (coated external building timbers out of ground contact). They cannot be used for any direct ground or water-contact applications.

The light organic solvent type preservatives (LOSPs) are being replaced increasingly by the micro-emulsion types, which are dispersed in water rather than solvent, reducing the emission of volatile organic compounds into the atmosphere. The LOSPs continue to be preferred for the particular treatment of decorative joinery components, minimising any raising of the wood grain. The micro-emulsion products are used extensively for the treatment of timber frame and truss material and general building timbers.

If these products are used to treat external joinery (HC3) then the timber should be subsequently protected with a maintained and appropriate surface coating. If the protective coating should fail the preservative performance may be compromised.

Both of these treatments leave the appearance of the timber virtually unchanged, although a colour can be added to the treatment for identification purposes, if required. The treatments also leave the dimensions of the timbers unchanged, unlike the water-borne high pressure treatments, which can swell the treated timbers.

LOSPs and water-borne micro-emulsion based preservatives are delivered using the double vacuum or Vac-Vac® process, summarised in Figure 4.4. Typical low pressure Vac-Vac® treatment schedules are detailed in Table 12 of BS 8417.

Coal-tar preservatives

Tar-oil or creosote-based preservatives have been used extensively in marine and fluvial environments and at one point were the pre-eminent preservatives used worldwide. A range of national specifications exists for creosote type preservatives and the specifier should initially refer to BS 8417. Since June 2003, the use of creosote-based wood preservatives in the European Union has been restricted to professional and industrial use only, owing to concerns that benzo-a-pyrene, one of the active compounds in creosote, is thought to be more liable to trigger cancer than was previously thought. However, there are also technical reasons why creosote treatments have declined in their use, one of which is the migration of the preservative to the wood surface under certain conditions.

The principal technical disadvantages of creosote are that it is not possible to apply paint or stain systems to creosote treated timber. The wood is greasy to the touch and is vulnerable to 'bleeding' – the creosote migrates to the surface of the timber and can be deposited on exposed skin, clothing and animals. In many recreational applications, this is undesirable and has led to its restriction. In such situations, if naturally durable timber is not specified, water-borne CCA or new generation type preservatives tend to be favoured.

Typical treatment schedules for timber to be treated under high pressure with tar-oil preservatives are summarised in Table 14 of BS 8417. Similarly to water-borne treatments, creosote is delivered into timber under high pressure using the 'full cell' process summarised earlier in Figure 4.3.

	Initial vacuum applied to remove air from the timber. Duration of the vacuum is determined by the treatability of the timber species.
	Vacuum held and treatment vessel flooded with preservative. Release of vacuum forces the preservative into the wood cells under atmospheric pressure. A low pressure may be applied for more resistant species or to achieve a higher specification, depending upon the end use of the timber.
	Second vacuum applied to evacuate excess preservative. Venting to atmospheric pressure drives surface preservative back into the timber and leaves the surface drip dry.

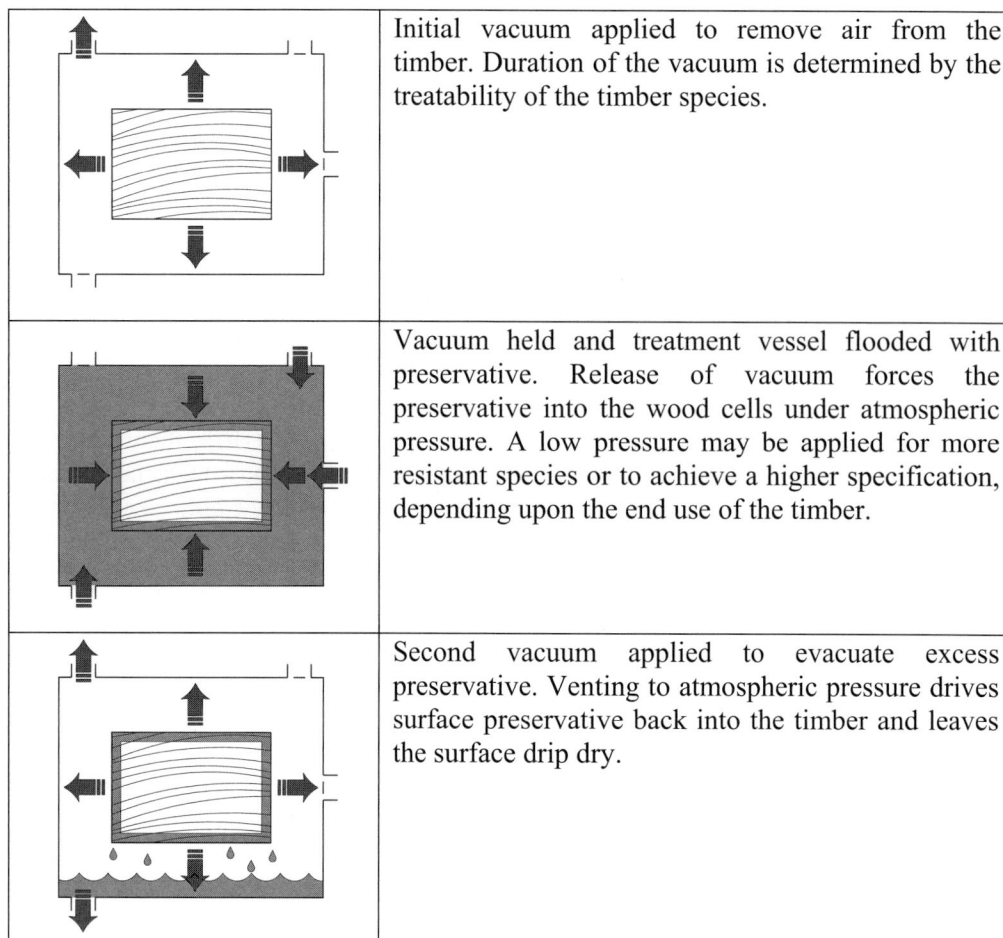

Figure 4.4. Summary of the Vac-Vac® low pressure treatment process (diagrams courtesy of Arch Timber Protection) Vac-Vac is a registered trademark of Arch Timber Protection

Sections 4.6.1 and 4.6.2 describe the current standards relevant to timber preservation and summarise the key points presented in BS 5268-5, BS 5589, BWPDA manual and BS 8417. The standards are comprehensive, but the information presented above can be summarised as a series of decision processes as in Table 4.9. The decision tree is based on the guidance given in Annex B of BS 8417 and the italicised text details the decision tree process for European redwood timber revetments serving in contact with fresh water.

Table 4.9. Stages of the decision process for using preservative treatment

Decision step	Action	Decision/ (*example*)
1	Identify timber component and service conditions	(*bank revetment for fresh water environment*)
2	Identify species to be used	Where a durable timber species is selected preservative treatment may not be necessary (*European redwood*)
3	Consider service factor and the need for preservative treatment	Depends upon the timber durability class, hazard class, service factors and desired service life (*Service factor D for European redwood in fresh water. Essential to use preservative treatment to minimise fungal hazard when non-durable timber is used as fresh water revetments*)
4	Select desired service life	Three desired service lives are identified: 15, 30 or 60 years *30 years*
5	Select service type	A preservative should be selected that is appropriate for the timber species serving in the specified hazard class. (*CCA type wood preservative conforming to BS 4072*)
6	Select preservative requirements	Select preservative performance requirements from Table 4 in BS 8417 (*Penetration class P9 Retention value 25 kg/m³*)

4.6.2. Understanding the standards for preservative treatment

In the UK, preservative treated timber for use permanently or intermittently in contact with sea or fresh water is usually specified in accordance with the guidance in either BS 5268-5, BS 5589, the British Wood Preserving and Damp-proofing Association (BWPDA, 1999) manual or, more currently, BS 8417. For all applications of preservative treated timber described in this manual, the preservatives must be applied under pressure.

BS 8417: 2003 *Preservation of timber – recommendations* is the most recent guidance document to be published in the UK. This standard describes the levels of preservative treatment to be achieved in terms of penetration and retention of preservative in the analytical zone of the timber which, previously, was not addressed in the process driven standards BS 5268-5 and BS 5589. Therefore, BS 8417 may be described as a target-driven standard. BS 8417 is a key document for specifiers which is supported by essential guidance detailed in the following documents.

1. BS EN 335-1: 1992 *Hazard classes of wood and wood-based products against biological attack – Classification of hazard classes*
2. BS EN 350-1: 1994 *Durability of wood and wood-based products. Natural durability of solid wood. Part 1. Guide to the principles of testing and classification of the natural durability of wood*
3. BS EN 350-2: 1994 *Durability of wood and wood-based products. Natural durability of solid wood. Part 2. Guide to natural durability and treatability of selected wood species of importance in Europe*
4. BS EN 599-1: 1997 *Durability of wood and wood-based products – Performance of preventive wood preservatives as determined by biological tests – specification according to hazard class*
5. BS 144: 1997 *Specification for coal tar preservatives*
6. BS 1282: 1999 *Wood preservatives – guidance on choice, use and application*
7. BS 4072: 1999 *Copper/chromium/arsenic preparations for wood preservation*
8. BS 5707: 1997 *Specification for preparations of wood preservatives in organic solvents*
9. BS 5268:-5: 1989 *Code of practice for the preservative treatment of structural timber*
10. BS 5589: 1989 *Standard code of practice for the preservation of timber*
11. BWPDA Manual: 1999.

The previously published BS 5268-5 and BS 5589 standards, and the BWPDPA manual (1999) provide detailed technical guidance on the type and use of various preservative treatments. The many schedules and variants of the basic procedures used in the UK are testament to the complexity and need for judicious use of effective timber preservation.

These documents are widely recognised throughout the wood preserving industry in the UK. BS 5268-5 describes the preservative treatment of timber used for structural purposes. In addition, BS 5268-5 classifies service factors based on safety and economic considerations related to wood destroying organisms and the need for preservative treatment of non-durable timber for a particular service category. This information is also presented in BS 8417 and is summarised in Table 4.10.

Table 4.10. Service factors based on safety and economic considerations

Consequence of failure of timber component	Need for preservative treatment (or specifying naturally durable timber species)	Service factor code
Negligible	Unnecessary	A
Where remedial treatment action of replacement is simple and preservation can be regarded as an insurance against the cost of repairs	Optional	B
Where remedial action or replacement would be difficult and expensive	Desirable	C
Where collapse of structures would constitute a serious danger to persons or property	Essential	D

The level of preservative treatment recommended depends not only on the risk of biological attack but also on the expected service life of the timber components. Desired service life values detailed in BS 8417 and its predecessors (BS 5268-5, BS 5589 and BWPDA manual) are not guarantees of performance but merely indications of the expectation against which the recommendations for timber treatment are drawn up. Moreover, timber preservation is not a panacea for poor design and construction. Durability by design is an essential factor that must be considered, and incorporating the principles of good design (see Section 7.3.1) can enhance the level of protection provided by the preservative treatment.

BS 5589 describes the preservative treatment schedules to be used for timber that is to be introduced to various hazardous environments and details preservative type, concentration of preservative solution where appropriate, timber treatability and treatment procedures. In short, BS 5589 can be described as a process driven standard based on previous experience and the long term track records of established wood preservatives such as creosote and CCA (copper/chromium/arsenic) containing preservatives. The BWPDA manual describes additional performance criteria called 'Penetration (P) Classes' for the penetration depth of preservatives into timber. These 'P' classes also incorporated into BS 8417 are discussed later in this section and are influenced by the permeability of the timber, in other words, the ease in which the timber can be treated.

4.6.3. Classification of the permeability of timber to preservative treatment

The life expectancy of non-durable or perishable timber components may be enhanced by pressure treatment with an appropriate wood preservative. The efficacy of the pressure treatment and the service applications of the timber will also be governed by the permeability or treatability of the timber to the preservative. BS 5589 and BS EN 350-2 describe four categories of permeability or treatability, listed in Table 4.11 below. BS 8417 simplifies these four permeability classes into either 'permeable wood' or 'resistant wood'.

Table 4.11. Classification of permeability classes

Treatability class BS 5589	BS EN 350	Description	Explanation	Examples
Permeable	1	Easy to treat	Easy to treat; sawn timber can be penetrated completely by pressure treatment without difficulty	European beech, pine sapwood
Moderately resistant	2	Moderately easy to treat	Fairly easy to treat; usually complete penetration is not possible, but after 2-3 hours pressure treatment, more than 6 mm lateral penetration (depth) can be reached in softwoods and a large proportion of hardwood vessels will be penetrated.	European elm, pine (heartwood), sitka spruce sapwood
Resistant	3	Difficult to treat	Difficult to treat; 3-4 hours pressure treatment may not result in more than 3-6 mm lateral penetration	Douglas fir, European larch, obeche, spruce heartwood
Extremely resistant	4	Extremely difficult to treat	Virtually impervious to treatment; little preservative absorbed even after 3-4 hours pressure treatment	Basralocus, greenheart, jarrah, European oak, teak

Note: Permeability rating for 'pines' are generalisations and the reader should refer to BS EN 350-2 for further details.

Where high preservative loadings are required to protect timber, only those species that are permeable or moderately permeable should be used. Deeper and more uniform preservative penetration into timber may be achieved by incising the wood. For example Douglas fir, although classified as being moderately durable against biodeterioration, has been used extensively in the marine environment. However, the sapwood of Douglas fir is moderately resistant to treatment and the heartwood is classified as resistant. Greater longevity can be achieved by incising the surface of the timber with sharp blades at regular intervals to improve the penetration of preservative fluids into the wood.

Table 4.12. Preservative treatment schedules as detailed in BS 5589

Timber species	CCA[1] treatment Fresh water (HC4) and salt water (M)			Creosote[2] by pressure process Fresh water (HC4)		Sea water (M)	
	Vacuum period (minutes)	Pressure period (minutes)	CCA conc. (g/L)	Min average net retention (kg/m^3)	Extended pressure period (minutes)	Min average net retention (kg/m^3)	Extended pressure period (minutes)
European redwood	60	180	50	160	180	200	180
Beech					120	400	180
Pitch pine					180	180	240
Elm					180	200	180
Douglas fir					180	180	240
Larch					120	200	180

1   CCA water-borne preservative should be applied by full cell process in compliance with BS 4072: Part 1. Vacuum of –0.8 bar applied followed by impregnation at 12.4 bar. N.B. CCA Treatments have been prohibited for use in marine environments from 30/06/2004.

2   Creosote preservative should comply with type 1 of BS 144: Part 1.

Where high populations of marine borer are known to be present, all sawn timber requiring preservative treatment, with the exception of elm, should be incised prior to treatment. Incising can also reduce the severity of splitting in service.

The timber treatments summarised in Tables 4.12-4.14 are intended to provide maximum protection to timber for use in fresh water and marine environments. The expected service life of the components is determined by the extent of wood decaying fungi and/or marine borers in the local environment.

To obtain a high, consistent standard of preservative treatment, consideration should be given to using the timbers in the round. The sapwood band is usually more permeable than heartwood and this should ensure a greater depth of preservative penetration and, also, greater preservative retention in the treated zones. Furthermore, there are mechanical advantages of using timber in the round, since poles are stronger than sawn timber of equivalent dimensions as there is less grain disruption in round timber. BS 8417 defines three desired service lives of 15, 30 and 60 years. The treatment schedule, preservative type, preservative concentration and permeability of the timber will influence the ability to achieve the desired service life.

Assuming good design and detailing preservative treated timber can, in practice, last far in excess of the desired service life expectancies stated in BS 8417. For all pressure treatments, all cross-cutting, mortising and drilling of timber components should be carried out before treatment. If this is not possible then any surface exposed by further matching/working after treatment should be liberally treated with an appropriate preservative to minimise the risk of attack.

Table 4.12 describes the treatment processes with water-borne CCA type preservative and creosote preservatives detailed in BS 5589 that should be followed in order to achieve the desired level of protection in service. These processes have evolved over many years of experience. The key difference between BS 5589, BS 5268-5 and the recently published BS 8417 is that BS 8417 (which will supercede BS 5589 and BS 5268-5) defines the level of penetration and retention of preservative required to protect timber in its service environment. Table 4.13 summarises the guidance given in BS 8417 for timber treated with CCA preservatives for use in brackish and fresh water environments and Table 4.14 details the guidance given in BS 144 for timber treated with coal-tar creosote containing preservatives. These tables describe the treatment processes for these two types of timber preservative which reflected their long established excellent track record for use in marine and fresh water environments.

Table 4.13. Treatment recommendations for CCA preservatives

| | Hazard class | Service factor | Desired service life | | | |
| | | | 15 years Permeable or resistant wood | | 30 years Permeable or resistant wood | |
			Penetration class	Retention	Penetration class	Retention
Timber in fresh water	4	D	P9	25 kg/m^3*	P9	25 kg/m^3*
Timber in salt water	5	D	Not permitted for use in marine environment (from 30/06/2004)			

* These values are interim critical values adjusted to take into consideration the industrial experience of the UK wood preserving industry and are an interpretation of the process guidance detailed on BS 5589 and BS 5268-5.

Table 4.14. Treatment recommendations for creosotes conforming to BS 144

| | Hazard class | Service factor | Desired service life | | | |
| | | | 15 years Permeable or resistant wood | | 30 years Permeable or resistant wood | |
			Penetration class	Retention	Penetration class	Retention
Timber in fresh water	4	D	P9*	320 kg/m^3*	P9*	350 kg/m^3*
Timber in salt water	5	D	P9*	320 kg/m^3*	P9*	350 kg/m^3*

* P + R values refer only to the analytical zone. The analytical zone is the area where preservative has been delivered into the timber.

To meet certain 'P' (preservative penetration) classes it may be necessary to incise the timber. The major implication of introducing target 'P' classes is that the specifier should have an understanding of the technical properties of the timber(s) to be treated. It may not be always possible to achieve the more onerous penetration classes required for preservative treatment in certain species of timber that are non-durable and resistant to treatment.

Analysis of species with permeable and moderately permeable sapwood can provide a good indication of whether the minimum sapwood retention values have been achieved and this is a requirement to establish if the desired treatment level specified in BS 8417 has been achieved. BS EN 351-1 sets out the performance criteria for the maximum permissible number of non-conforming samples, assessed as part of a quality assurance programme within the industrial pre-treatment plant, based on whether the timber is classified as 'permeable' or 'resistant' to treatment. Many treatment plants are automated and many preservers issue treating certificates to assure end users that the timber has been treated in accordance with the relevant schedule detailed in the end users' specification.

Qualitative analysis of treated timber may be undertaken to determine that the required preservative loadings have been applied. These tests are detailed in Parts 1 and 2 of BS 5666: *Methods of analysis of wood preservatives and treated timber.*

Table 4.15 is based on data presented in both the BWPDA manual and BS 8417 and summarises the information detailing the penetration classes and the analytical zones for determination of preservative retention in accordance with BS EN 351-1.

Table 4.15. Penetration classes and definition of the analytical zones

Penetration class	Penetration requirement	Analytical zone
P1	None	3 mm from lateral face
P2	Minimum 3 mm lateral and 40 mm axial penetration into sapwood	3 mm lateral into sapwood
P3	Minimum 4 mm lateral penetration into sapwood	4 mm lateral into sapwood
P4	Minimum 6 mm lateral penetration into sapwood	6 mm lateral into sapwood
P5	Minimum 6 mm lateral and 50 mm axial penetration into sapwood	6 mm lateral into sapwood
P6	Minimum 12 mm lateral penetration into sapwood	12 mm lateral into sapwood
P7	Roundwood only: Minimum 200 mm lateral penetration into sapwood	20 mm into sapwood
P8	Full sapwood penetration	Sapwood
P9	Full sapwood and minimum 6 mm penetration into heartwood	Sapwood and 6 mm into exposed heartwood

In addition to defining the performance characteristics for preservative penetration and retention classes, BS 8417 also sets out the limits for what are known as critical values (CVs). Critical values may be defined as the minimum concentration of preservative required to inhibit the activity of wood-destroying organisms (fungi and/or insects) in treated timber.

CVs are applicable to new timber preservatives which have not been in commercial use sufficiently long enough for the relevant commercial and field experience to have been amassed. For example, CCA preservatives have a proven track record in use of about 70 years and the track record for creosote is even longer. Therefore CV limits are based on laboratory experience, not practical experience. With reference to new preservatives for use in fresh water and marine environments, for a service life of 15 years, the preservative should achieve the same penetration class (P9) as for more traditional treatments (CCA or creosote) and the CV should be the same as that achieved in laboratory trials. However, should a service life of 30 or years be desired, the CV derived from laboratory trials should be multiplied by a factor of 1.5 or up to 2.5, depending upon the expected service life and treatability of the specified timber. See Table 4.16.

Table 4.16. Treatment recommendations for preservatives for which an appropriate CV value as described in BS EN 599-1 is available

| | Hazard class | Service factor | Desired service life | | | | | |
| | | | 15 years | | 30 years | | 60 years | |
			Permeable and resistant wood		Permeable and resistant wood		Permeable and resistant wood	
Timber in fresh water	4	D	P9	CV1	P9	CV1.5	P9	2.0
Timber in salt water	5	D	P9	CV1	P9	CV1.75	P9	2.5

As stated earlier, industrial timber preservation provides a protective envelope around the component. In environments where there is a high risk of abrasion and/or wear, the protective envelope may be eroded and expose vulnerable, unprotected timber.

Where preservative treatment is not desirable or practicable, owing to a high risk of abrasion in service or strength-limiting factors, the designer may specify heartwood of timber species that are considered to be naturally durable and, where relevant, resistant to attack by marine borers. Table 4.17 details a number of appropriate timber species. The information is principally taken from BS 5589.

Table 4.17. Suitability of timber (heartwood) for use in sea and fresh water

Species suitable for use untreated in sea water (M)	Species suitable for use untreated in fresh water (4)	Species suitable, if treated, for use in sea water (M) and fresh water (4)
Andaman padauk	All timber that is rated as very durable (VD) suitable for this application	American pitch pine
Basralocus		Caribbean pitch pine
Ekki		Douglas fir
Greenheart ♦		Elm
Iroko		European larch
Jarrah		European redwood
Kapur		Japanese larch
Okan		
Opepe		
African padauk*		
Belian*		
Iroko*		
Muninga*		
Pyinkado*		
Red louro*		
Teak*		

*   Species of timber detailed in BWPDA manual (1999).
♦   Care should be taken with greenheart as the sapwood is not as easily distinguished from the heartwood as within other species such as ekki, for example.

The list of species given in Table 4.17 is by no means exhaustive and other species, especially lesser known species, may be suitable for marine and fresh water environments.

4.6.4. *Environmental considerations and European Union legislation*

The Biocidal Products Directive now being introduced will standardize approval regimes for all biocidal products, including wood preservatives, across the European Community. The comprehensive controls already in place in the UK under the Control of Pesticides Regulations will result in this directive having a minimal impact in the UK. However, the Directive does mean that approval for all existing biocidal products will have to be reviewed. Following the review period, the wood preservation industry expects stricter controls to be implemented. These controls may take the form of stricter voluntary regulation driven by the industry itself.

Section 4.6.3 briefly described water-borne preservative such as those containing salts of copper, chromium and arsenic (CCA). The use of CCA containing preservatives has been the subject of intense debate throughout Europe. Consequently, the Scientific Committee on Toxicity, Ecotoxicity and the Environment (CSTEE) has undergone a protracted consultation process to assess the risks to health and to the environment of arsenic in wood preservatives and of the effects of further restrictions on its marketing and use.

In November 2002, the amendment to Council Directive 1976/769/EEC restricted the use of timber treated with CCA solutions in marine waters. This restriction came into force on 30th June 2004. However, CCA treated timber may still be used as constructional timber in fresh water areas and brackish waters. CCA treated timber may also be used as structural timber in earth-retaining structures, public and agricultural buildings, office buildings, industrial premises and bridges and bridgework. In certain situations, this restriction will force the designer to design with tropical hardwoods or concrete and steel.

Timber preservatives not containing arsenic but suitable for use in marine and fresh water environments are available within the EU. However, these formulations tend to be more expensive, and confidence in the service life of the treated timber is not fully established.

As noted in Section 4.6.1, the use of creosote-based wood preservatives in the European Union was restricted from June 2003 for health and technical reasons. Certain uses of creosote-treated wood are still permitted provided the treatment has been carried out in an industrial installation or by professionals under specified conditions.

4.7. PROPERTIES OF RECYCLED TIMBER

Recycled timber can provide a valuable source of material. Old, redundant structures may be viewed as 'industrial forests' and their yield will depend entirely upon the quality of the reclaimed timber and its condition. Reclaimed timber (see Figure 4.5) should be set apart from 'new' timber and certain guidelines should be followed. The first question that needs answering is: 'What species am I reclaiming?'. It is critical that the right species for the job is selected. By identifying the species, the salvager can judge whether it is worth recycling the timber and for what markets or use this material is appropriate.

This section briefly describes the pertinent points to consider when reclaiming timber. Potential sources of reclaimed timbers include posts and beams from demolished warehouses, pilings and beams from dismantled piers and planks, and piles from dismantled groynes. For all reclaimed timber, it is important that the members are stripped of all ironmongery prior to reprocessing otherwise costly and time-consuming damage can be caused to the machinery.

Figure 4.5. Reclaimed hardwood piles awaiting sorting, machining and grading (courtesy John Williams)

Generally, the timbers should be machined so that worn, weathered and damaged surfaces are removed to expose the underlying timber. This will inevitably result in a loss of section. The freshly exposed surfaces are now in a presentable condition to enable the timber to be graded to the appropriate visual or machine-grading standard (Figure 4.6).

In addition to naturally occurring strength-reducing features such as knots and slope of grain, the timber may have been significantly affected by man-made defects such as holes for fixings and notches. In practical terms, where high volumes of timber are being recycled, heavily notched timbers should be machined so that these areas are removed. Where notches are infrequent, they may be considered as 'knots' and the grading rules for knots may be applied. Holes for fixings can also be considered as knots.

Figure 4.6. Sorted and part machined reclaimed hardwood timber (courtesy John Williams)

The timbers should be assessed for any signs of biodeterioration such as fungal decay and marine borer attack. Typically, survey techniques include decay detection drilling and hammer soundings.

Mechanical damage may affect the strength of recycled timber significantly. If there are visible signs of impact, there may be a risk of compression failure within the timber. If compression failure is suspected, the timber should be rejected and cross-cut to exclude the impact area. Large section timbers such as balau, for example, may have been cut in such a manner that the member contains boxed heart. Boxed heart arises when the weaker, less durable timber comprising the pith in the centre of the tree is protected by stronger durable material. Examining the end grain usually identifies timbers that exhibit boxed heart. If the pith is visible, then the member contains boxed heart. Provided that the boxed heart is still armoured by older mature timber after cleaning, the reduced section may be used. However, it may not be advisable to place the recycled member in a critical part of any structure.

If the reduction of the section results in exposure of the pith along the length of the member then it should be rejected. Generally, most large section timbers that have boxed heart are easily identifiable by heart checks at the end grain. Heart checks are characteristic of old, large section timbers that have dried out. Research in the USA has shown that heart checks have little effect on recycled timbers used as columns but they do cause a significant reduction in the modulus of rupture when used as beams.

There are other factors to consider. Large section timbers that have been serving in dry conditions will be well seasoned and, as such, will be dimensionally stable. In situations where little timber movement is expected, recycled timbers may have significant advantage over large sections of green timber.

The moisture content of the timber is vital. For example, should water logged timber members be broken-down to smaller section members, it is inevitable that as they dry out they may experience some distortion. This may be minimised by weighing down stacks of the broken down members to restrain the timber. The potential effects of timber movement should be considered when using recycled materials.

The properties provided within Appendix 1 are generally applicable to recently produced timber. Whilst recycled timber of the same species is likely to have similar characteristics it may, depending on age and exposure, be more brittle and more difficult to work. In addition, repeated loading may have caused damage which, although difficult to detect, may have weakened the timber. If there is doubt as to the properties or suitability of recycled timber a qualified timber strength grader should be consulted.

Recycled timber deemed to be unsuitable for further structural work should be made available for other uses (including landscape or decorative features) rather than disposal to landfill sites.

4.8. STRUCTURAL TIMBER COMPOSITES

Although timber provides high specific strength and stiffness, its natural variability has, traditionally, challenged the engineer. Through efficient strength-grading procedures and judicious selection of material, highly efficient structural composite materials have been developed. The major advantages of structural timber composites are that large dimensions are available and that higher strength characteristics can be achieved compared with those of the raw material. Essentially, the manufacture of these composites involves transforming relatively low value raw material into high value engineered products.

The structural timber composites described below are suitable for external use as defined in Hazard class 3 with the appropriate preservative treatment, if manufactured from non durable materials. Their suitability for use in Hazard classes 4 and 5 should be checked with the manufacturer and whilst there are only limited examples of marine applications identified at the time of writing, there is thought to be considerable potential for their further development and application to coastal and river structures.

4.8.1. Glulam

The definition of glued laminated timber (Glulam) in European Standards is that it is manufactured from laminates of timber glued together with the grain in the laminates running essentially parallel. The laminates in Glulam are normally sawn sections which are planed to a smooth surface before gluing.

The laminates are butt jointed using finger jointing techniques so the overall dimensions of the finished product are determined by the manufacturing premises and transportation facilities. The laminates are bonded together to form a member that acts as a single structural unit. Where appropriate the laminates are preservative-treated prior to manufacture. The manufacture of glulam is strictly controlled by a series of British and European standards which focus on batch testing of the product for mechanical strength and glue line integrity.

The manufacture of glulam with finished lamination thicknesses of not more than 45 mm is covered by BS EN 386: *Glued laminated timber – Performance requirements and minimum production requirements*. This standard replaces BS 4169: 1988 *Specification for manufacture of glued laminated timber structural members*. The BS EN 386 requirements apply to structural members used in Service classes 1 and 2. For members used in Service class 3, special precautions need to be taken as, for example, in the use of weather-resistant adhesives as defined in BS EN 301: *Adhesives, phenolic and aminoplastic, for load-bearing timber structures. Classification and performance requirements*.

The general aim of BS EN 386 is to ensure that reliable and durable bonding is achieved consistently. To this end it lays down requirements relating to premises, staff and equipment as well as technical aspects such as the moisture content of the laminates at the time of gluing, the orientation and lay-up of the laminates, cramping pressures, curing time and temperature for the glues. Emphasis is laid on quality control, with routine daily testing of finger joints and laminate gluelines being required as part of the production process. The manufacturer is required to keep detailed records of every production run. External control through independent third party quality certification to oversee the quality of production is required.

Glulam (Figure 4.7) is used extensively in the UK and the European mainland and has been used in a wide variety of applications ranging from high profile 'flagship' developments to industrial premises. It is available as 'off the shelf' prefabricated components or specialised sections can be manufactured to order, for example, curved sections. Glulam may also be manufactured in the vertical form.

Figure 4.7. Glulam (courtesy TRADA Technology)

Glulam has a number of advantages over solid timber components. Glulam is available in large cross sections and lengths at a controlled moisture content. It can be manufactured in curved, straight or cambered shapes. The strength of glulam is determined by the grade and species of timber, both hardwoods and softwoods, used in its manufacture. This makes glulam a very versatile material that can be used in the construction of many structural forms.

The laminating process disperses the natural strength-reducing features such as knots and sloping grain throughout the member. This produces an engineered product that has higher strength and improved dimensional stability. Most commercially available glulam comprises laminates from a single species and grade. Most 'off the shelf' glulam is manufactured from European whitewood which is resistant to preservative treatment. Therefore, for external uses, glulam manufactured from pre-treated European redwood or from naturally durable hardwoods such as iroko, which has an established track record, is preferred. Other lesser known hardwoods would have to be assessed by a competent authority.

The advantages of such an engineered structural composite is that strength-reducing features such as knots may be eliminated by cross-cutting. In addition, a balance between maximising strength with attention to efficient use of materials may be achieved by sandwiching laminates with strength-reducing features between 'defect' free laminates so that the higher quality, stronger laminates are positioned in places of maximum stress. Also, in the lower and mid-range strength classes, laminated timber has a higher characteristic bending strength than the timber from which it is made because defects are eliminated or spread through the glulam section.

Glulam can be either homogenous, where all the laminates are the same strength class of timber, or combined, where the outer laminates (one-sixth of the depth on both sides) are a higher strength class than the inner ones.

For mechanically fastened joints in glulam, as in solid timber, the formulae in Eurocode 5 require the characteristic density of the material rather than its strength class. Lamination averages out this value, so the load-carrying capacity of mechanically fastened joints calculated from Eurocode 5 will, in many cases, be increased by laminating the timber. Eurocode 5 also gives advantageous rules for plain dowelled joints which are useful in glulam structures.

4.8.2. Mechlam

Mechanically laminated timber consists of members which are straight or slightly curved and which are formed from laminations connected together by dowel-type fasteners. The use of mechanically laminated timber for construction in Europe was well established by the start of the nineteenth century and comprised thick-bolted or rod-connected laminations. Connections between adjacent layers are normally made with stainless steel dowels force driven through pre-drilled holes in the laminations which have smaller diameters than the dowels. This produces a tight fit at the interface.

4.8.3. Steel and timber composites – prestressed decking plates

Laminated decking plates are made up of individual laminations which are held together by nailing or adhesive bonding. In the case of prestressed plates there is, in addition, a permanent lateral pressure which guarantees continued friction contact between the faces of the laminations and any unbonded adjacent faces.

Stress-laminating may be defined as a permanent effect due to controlled forces imposed upon the structure. Timber plates are normally prestressed by means of steel bars or tendons. Care must be taken when designing prestressed timber plate systems as incorrect moisture content specification can lead to a loss in prestressing forces as a result of timber shrinkage. Of course, the steel should be corrosion resistant. Steel may also be substituted for glass reinforced fibre rods. Prestressing rods are usually bonded into the structure which increases corrosion resistance and results in good load-carrying capacity.

Steel reinforced timber elements serve in high duty environments although localised reinforcing may be employed to repair/upgrade structures. Reinforcement is especially worth considering where elemental design is restricted through the relatively low tensile strength of timber transverse to the grain direction. Steel reinforcement may also be used to increase resistance to shear.

4.8.4. Laminated veneer lumber

Laminated veneer lumber (LVL) consists of veneers, typically of an initial thickness of 3.2 mm which are peeled from logs and bonded together, generally parallel to one another. LVL offers similar advantages to glulam. It attains even more strength since the laminations are thinner and timber growth features have greater dispersion throughout the member. Similarly to glulam, LVL offers properties which may be used for many load-bearing forms such as principal beams for bridges, for example. LVL is only manufactured in long straight boards which are subsequently rip-sawn into long members. Therefore, although it is possible to order LVL cut into tapered or curved forms, it is more economical to use it in straight sections.

The manufacturing process begins by soaking debarked logs in hot water for 24 hours. They are then rotary peeled to produce veneers, typically 3-4 mm thick which are clipped into sheets 2 m wide. After drying, phenol formaldehyde adhesive is applied to the veneers which are normally laid with the grain running parallel to form a mat of desired thickness. The joints are staggered vertically to minimise their effect on the strength of the LVL. The mat of veneers is then pressed at high temperature and then cross-cut and rip-sawn to the required dimensions.

As with glulam manufacture, quality control during manufacture includes regular testing of the quality of the glue bond and of the mechanical properties such as bending strength of the finished material.

4.8.5. Parallel strand lumber

Parallel strand lumber (PSL) is manufactured in the USA mainly from Douglas fir and Southern yellow pine. As in the manufacture of LVL, logs are peeled into veneers. The sheets of veneers are then clipped into strands up to 2400 mm in length and 2-3 mm in thickness. A phenol formaldehyde adhesive is used which is mixed with wax to enhance the dimensional stability of the finished product. The adhesive is applied to the oriented strands which are fed into a rotary belt press and cured under pressure by microwave heating. PSL emerges as a continuous billet which is then cut and trimmed to the required dimensions.

4.8.6. Laminated strand lumber

Laminated strand lumber (LSL) is also manufactured in the USA. LSL is manufactured from low value aspen strands up to 300 mm in length and 30 mm in width. After drying the chips are coated with a polyurethane adhesive and the chips are oriented in a direction parallel to the panel length. The resulting mat is heated and pressed to produce a finished panel which is cut to the required sections.

4.8.7. Advanced wood composites

In the USA, the research and development of wood–plastic composites is receiving much attention. This has largely been borne out of the combination of the restrictions of using pressure treated softwood (creosote and CCA) and a growing reluctance to specify durable tropical hardwoods. By far the biggest driver is the classification of redundant pressure treated components as hazardous waste. The cost implications of disposal are prohibitive and, historically, tend not to have been considered as part of the life cycle cost analysis of a structure.

In addition, new shore front structures require large service loads and designers are steered towards steel and concrete solutions. However, there is market opportunity for wood–plastic composites and the US Navy is working closely with researchers to develop and use carbon reinforced engineered thermoplastic timber composites. Typically, these composites consist of engineered particles, fibres or flour. In addition, carbon fibre reinforced composites are being assessed, such as glulam with carbon fibre reinforcements at the interfaces.

An advanced wood–plastic composite (WPC) describes the material formed when wood material is impregnated with a monomer. The aim of developing WPCs is to combine the tensile strength of wood fibres with the abrasion resistance, dimensional stability, hardness and incompressibility of the polymeric materials used in the manufacture of the WPC.

4.8.8. The use of structural composites in the marine and fresh water environments

There are issues which the designer should be aware of when considering the use of structural timber composites in a submerged environment. Mechanically or glued laminated timber has a track record of being used in the marine environment. There are many instances in boat building where the keel has been made of glued laminated timber.

The adhesives used in the manufacture of glulam and other structural timber composites tend to be phenol resorcinol-formaldehyde (PRF). These adhesives have a proven track record and are used in the manufacture of external and marine grade plywood.

Provided the designer chooses a naturally durable timber, there should be little difficulty in designing a mechlam or glulam structure for service in the marine environment, although there are few instances where this has been carried out. The principal advantage of considering glulam as a possible solution is that smaller section sizes of timber may be utilised which can widen the choice of candidate species.

If choosing to use a non-durable timber, timber preservation needs to be considered. This can pose considerable problems. Firstly, the expected restrictions in the use of CCA type preservatives on the marine environment restricts the choice of preservative. Secondly, abrasion of the components can breach the preservative envelope which can shorten the expected service life of the structure. The feasibility of using preservative treated components may also be restricted by the size of the chamber available to accept the finished component. Treatment of individual laminates is not available from all glulam manufacturers.

The use of materials such as laminated veneer lumber (LVL) and parallel strand lumber (PSL), for example, should be limited to Hazard class 3 environments. The manufacturer of the product should always be consulted by the designer to check that the manufacturer is confident that their product is fit for the intended purpose.

In most external structural applications the service life of structural timber composites can be virtually unlimited. The use of modern external class adhesives ensures excellent glueline durability. The main threat to the service life of such components out of ground and water contact is a rise in moisture content. However, good design coupled with good maintenance procedures can minimise the risk of deterioration.

In summary, mechlam and glulam timber may be used in marine and fresh water environments provided they are manufactured out of suitably durable materials or preservative treated materials that conform to European legislation. Other structural timber composite such as LVL and PSL should only be used in environments where the service risks are no greater than those defined by Hazard class 3.

5. Responsible procurement of timber

The selection and specification of appropriate materials has a significant influence on the environmental impact and sustainability of schemes as well as the performance and durability of the timber structures.

It is essential that the inherent advantages of timber are not negated by irresponsible procurement practices. This will necessitate a more detailed assessment of the requirements of a particular scheme and the properties (including environmental provenance) of timbers used in any solution. Whilst selection will include assessment of engineering and economic issues in addition to environmental considerations, particular attention will be needed to ensure that the use of tropical hardwoods, especially those which are not demonstrably from legal and sustainable sources, is clearly justified. However, care must be taken to ensure that timber is not so difficult to procure that it is substituted by more environmentally damaging materials (see Chapter 2).

5.1. ENVIRONMENTAL CONSEQUENCES OF TIMBER USE

Tropical hardwoods have properties which make them attractive for coastal and river structures, however in recent years there has been increasing public concern relating to environmental damage caused by logging of forests.

5.1.1. Forest loss and degradation

The complex biological structure within natural tropical forests makes them particularly vulnerable to over-exploitation which can result in irreversible deforestation. The Global Forest Resources Assessment 2000 (FAO, 2001) prepared by the United Nations Food and Agricultural Organization estimates that the 14.6 million hectares of natural forest are lost each year and a further 1.5 million hectares per year of natural forests were converted to forest plantations. This is to some extent offset by the expansion of natural forests and forest plantations (at a rate of 5.2 million hectares per year) and the rate of deforestation in the 1990s was slower than that of the 1980s. However, most of the forest losses were in the tropics and it is widely accepted that the startlingly rapid rate of tropical forest destruction experienced in recent years is unsustainable.

The loss of forest (Figure 5.1) has significant local impact (including reduced water quality and availability of forest resources, falling water tables and increased risk of drought, landslides and flash floods) as well as more global impacts such as reduced biodiversity and air quality, global warming and the continuity of supply of timber. The loss of natural forest is not directly attributable to any single activity, but a combination of urban development, logging and land clearance for agriculture, plantations, infrastructure and mining. The growth in demand for edible oils (particularly palm and soy) is thought to have contributed to the conversion of 300 million hectares of tropical forest to plantations over the last 20 years (WWF, 2002a). Whilst it is possible for plantations to be managed in an environmentally sensitive way, many place great stress on the local environment and society as they are managed to maximise the short term yield, often through the intensive use of irrigation and agrochemicals.

Figure 5.1. Loss of natural forest (courtesy Richard Copas)

Illegal logging is another major contributor to forest degradation and loss, but due to the nature of the activity it is very difficult to quantify. Illegal logging deprives governments of revenue which, in turn, has significant impact on social provision such as education, sanitation and healthcare. It is seldom undertaken with any regard to sustainable forest management and may lead to increased agricultural conversion, mining, poaching, human settlement and forest fires. An initiative by the G8 governments to reduce illegal logging is summarised in Box 5.1.

Box 5.1. Illegal logging

Following a meeting of G8 finance and foreign ministers in May 1998 a G8 Action Programme on Forests was announced (G8, 1998). This included the following on illegal logging:

"Illegal logging robs national and subnational governments, forest owners and local communities of significant revenues and benefits, damages forest ecosystems, distorts timber markets and forest resource assessments and acts as a disincentive to sustainable forest management ..."

The G8 members gave a commitment to:

- encourage the sharing of information and assessments as a basis for developing countermeasures
- improve economic information and market transparency
- assess the effectiveness of internal measures to control illegal logging
- implement obligations under international agreements to combat bribery and corruption
- work with interested partner countries to develop their capacity to assess illegal logging and develop and implement counter measures.

More recently, the UK Government commitment to tackling illegal logging was reiterated: *"The UK Government has taken the lead in tackling illegal logging. Our forests are the single most important stabilising feature of the land's surface. The commitment that the Government made to purchase timber products only from legally logged, sustainably managed sources is a significant step. It is also a challenging target. This commitment has been acknowledged across Government, and an inter-departmental buyers group has been established to advise and monitor on performance."* (DEFRA, 2001).

Recent analysis (Report *Controlling the Trade in Illegal Logging*, Royal Institute of International Affairs (RIIA) 2002; WWF Report 2002 on Illegal log procurement by governments *Timber Footprint of the G8 and China*) suggests that better information on the extent of illegal logging is now available and much more could be done to tackle illegal logging. At an EU level, the Forest Law Enforcement, Governance and Trade (FLEGT) initiative suggests that tighter controls on timber imports should be under consideration to help prevent forest crime. *"Illegal logging and forest crimes play a significant part in the ongoing extensive destruction of the tropical forest (an area of about five times the size of Belgium is lost every year), with consequent loss of government revenues, estimated at up to $15 billion per year globally, and of funds for social services and attempts to reduce poverty. So there is now general agreement that measures should be taken to prevent the import into the EU of illegally sourced timber".* FLEGT Report 2002.

Whilst unsustainable logging for export does contribute to the loss and degradation of forests, only a relatively small proportion of the harvested timber is actually exported. In many areas much of the timber is used for firewood, local construction or wood products such as paper. Nevertheless, the responsible specification and purchase of timber can have a significant impact in discouraging illegal and unsustainable practices.

5.1.2. The case for using timber

Timber is renewable and is an environmentally responsible material option if recycled or if known to come from sustainably managed forests. When this is the case, there are very strong environmental arguments in its favour – with relatively low energy required for its production and (apart from transport and processing) being virtually 'carbon neutral'. Indeed, all forests sequester carbon from the atmosphere both in living biomass and in forest soils. Living forests also have important functions in the regulation of other cyclical processes important to the earth's climate – particularly air quality and the water cycle.

5.1.3. Sustainable forest management

Most types of forest can be managed sustainably, that is to say, harvested periodically for timber and other products in such a way that their productive benefits for future generations are not compromised. It can be argued that sustainable management and exploitation of forests provides an economic *raison d'être* which discourages conversion for agricultural or plantation use. Many European forests have been managed for hundreds of years and such management is not necessarily incompatible with high levels of biological diversity, rather careful long term management can enhance the biological diversity of forests.

Whilst the term 'sustainable forest management' will mean different things to different people, there is a consensus amongst many forestry professionals that it is a 'multi-dimensional balancing act' (Upton and Bass, 1995) involving economic and social as well as environmental interests. This is succinctly expressed by principle 1b of the Non-Legally Binding Statement of Forest Principles (Rio, 1992): *"Forest resources and forest lands should be sustainably managed to meet the social, economic, ecological, cultural and spiritual needs of present and future generations."*

5.2.   PROCUREMENT POLICY AND ISSUES

There is an urgent and widely recognised need to encourage and implement sustainable forest management worldwide. This is evident from international agreements, government commitments and intense lobbying from environmental non-governmental organisations. There are strong drivers for both public and private entities to adopt procurement policies that will favour legally harvested wood from sustainably managed forest. Government and trade measures to protect forests are of diminished effectiveness in a market where stolen, illegally harvested or plundered wood is freely available and frequently much cheaper than legitimately produced timber. Any procurement procedure which does not incorporate environmental considerations will tend to favour decisions based only on price and thus encourage illegal logging and exploitative forestry.

5.2.1. Governmental policy and agreements

Concern at governmental level to encourage and implement sustainable forest management worldwide has manifested itself in several international agreements or forest protection processes as shown in Box 5.2.

Box 5.2.   International agreements on timber trade and sustainable forest management

- Agenda 21 of the Rio Earth Summit of 1992 – particularly Chapter 11 The Non-Legally Binding Statement of Forest Principles – Rio, 1992
- International Tropical Timber Agreement (ITTA) 1994
- The Helsinki Accord 1996
- The Montreal Accord and the Santiago Declaration 1995
- The UNCED agreements, including UNCBD, the International Panel on Forests (IPF), the International Forum on Forests (IFF), CITES (The Convention on International Trade in Endangered Species of Wild Fauna and Flora), and the Anti-Bribery Convention
- OECD and G8 statements.

The UK government has reiterated its policy on several occasions in recent years, perhaps most clearly in July 2000 when the Environment Minister stated

" ... current voluntary guidance on environmental issues in timber procurement will become a binding commitment on all central government departments and agencies actively to seek to buy timber and timber products from sustainable and legal sources, for example, those identified under independent certification schemes such as that operated by the Forestry Stewardship Council." (excerpt of speech by Michael Meacher, House of Commons 28th July 2000).

A recent report by the Environmental Audit Committee suggests that whilst the UK government policy is commendable, implementation by government departments and agencies is not yet widespread:

"Not only does the government have a responsibility to lead by example in environmentally sound timber procurement practices; it also has, through its buying power, the potential to change the nature of the timber markets through the procurement decisions that it makes ... While Government rhetoric has been laudable, we see no systematic or even anecdotal evidence of any significant change in the pattern of timber procurement since July 2000 ... " (House of Commons, 2002).

The Dutch government has taken a significant initiative in supporting sustainable forest management, not least through the minimum requirements shown in Box 5.3.

The UK, Dutch and other governments are also assisting producing nations, including Indonesia, under bilateral agreements.

Box 5.3.   Dutch minimum requirements

(Dutch Ministry for Nature Management, 1997)– Bold type from the original text
Minimum requirement 1
Local **forest management should demonstrably and sufficiently show that attention has been paid to the integrity of** ecological **functions and the continuity of** socioeconomic and **sociocultural functions of the forest on the basis of the relevant criteria and indicators.**

Minimum requirement 2
Local forest managers should have an adequate management system

Minimum requirement 3
The certifying body is independent, demonstrably complies with international regulations as to organisation and procedures followed and has the necessary forest management expertise.

5.2.2. Procurement based on understanding

In a world where diminishing and important natural resources need to be conserved, responsible procurement decisions can only be made with an understanding of environmental considerations.   The consequences of ignoring this can be considerable, impacting on both the project and the client as well as the wider environment, as is shown in Table 5.1.   The key to minimising controversy for specifiers is to be able to demonstrate that a logical and responsible procurement process is followed and that environmental considerations are included at all stages during the scheme development.   At the most basic level this requires that timber being procured can be proven to come from forests that are being legally harvested (see Table 5.2).   However, legality alone does not ensure sustainable yield production, so it is also highly desirable that timber comes from forests that are being 'sustainably managed' or at least managed in a way that harvesting is not overly detrimental and progress is being made towards sustainable management. Ultimately, all timber (both temperate and tropical) should come from legal, sustainable, managed forestry.

The UK currently imports about 700 000 m^{3}of tropical hardwoods per year (Environmental Audit Committee, 2002) and whilst only a relatively small proportion of this is thought to be used in coastal and river works, these applications form a significant market for a number of species such as greenheart (from Guyana) and ekki (or azobe from West and Central Africa).   Although the trade in tropical hardwoods has been the main focus of controversy, environmental and social problems resulting from the loss and degradation of forest areas are not limited to tropical regions. Responsible specification and procurement of temperate soft and hardwoods is also needed.

Table 5.1. Possible consequences of ignoring environmental issues in timber procurement

Public Sector	Private Sector
• Inefficient or inappropriate use of timber • Public criticism and delay to scheme approval / construction due to review of timber selection • Rejection of timber selection / specification by client due to non-conformity with environmental accreditation (e.g. BS EN ISO 14001), policy or reporting requirements • Legal challenges delaying or preventing scheme construction	• Inefficient or inappropriate use of timber • Concern by potential investors / clients due to increasing environmental awareness • Non-conformity with the requirements of environmental accreditation (e.g. BS EN ISO 14001) or other commitments (e.g. WWF 95+ group) • Public criticism and damage to reputation
Legal and sustainable procurement is an obligation upon government, government agencies and local authorities under Agenda 21 and other agreements.	*Private clients are under no legal obligation to observe environmental criteria. However, this does not mean that they might not be affected by consequences of uninformed decision-making.*

5.2.3. *Relevant guidance*

A recently published report (Procurement of timber products from 'legal & sustainable sources' by Government & its executive agencies, ERM, 2002) is suggesting a basis for public sector timber procurement in the UK. This report addresses appropriate sustainability criteria and recognises that it is necessary to adopt a progressive approach to eliminating illegal and unsustainable timber.

The preference for legal and sustainable timber is consistent with government policy and whilst it is not mandatory for private sector organisations it should be considered best practice. In order to identify whether a timber is legal and sustainable it is essential to have verification of where it has come from. This requires the 'chain of custody' to be recorded and possibly verified through independent auditing.

It is also important to realise that not all definitions of legality and sustainability are consistent. For the purposes of the guidance provided in this document, the definitions based on those developed for the UK Government (ERM, 2002) have been adopted (Table 5.2).

It is hoped that this will allow sufficient flexibility to meet particular technical requirements, whilst encouraging sustainable forest management and ensuring minimum standards. It is recognised, however, that assessing different types of evidence will be outside the competence of most individuals and it is proposed that this is undertaken by a 'central point of expertise' within the government.

The report from ERM, as well as the Dutch 'minimum requirements' (see Box 5.3) identify independent certification as the means of first choice by which both governments are able to ensure that they meet their environmental commitments to sustainable forests.

5.3. FRAMEWORK FOR RESPONSIBLE PROCUREMENT

In order to provide the required performance with minimal detrimental impact on the environment, the selection and specification of timber must balance requirements for technical performance with the provenance, sustainability, cost and availability of the timber(s). A framework enabling this is presented in Figure 5.2. The framework demonstrates the iterative nature of the design and timber selection process, and intentionally excludes cost criteria. Whilst there may be a cost premium for certified timber, it is likely to form a relatively small element of the overall scheme value.

Where appropriately certified timber is readily available the procurement process should be relatively simple. Where this is not the case, it must be recognised that a degree of compromise is required.

It is likely to be some time before certified sources for the wide range of timber used in coastal and river works will be available in sufficient quantities. In the interim this framework is intended to support sustainable forest management practices, wherever they are occurring. It should be noted that, given the current harvesting regime in many forests, timber which is offered for sale without supporting documentation (see Table 5.2 for examples) could well be illegal or unsustainably obtained from the forest. It is important that potential suppliers are aware of requirements for documentary evidence at the start of the procurement process, as post-contract investigation could prove very expensive.

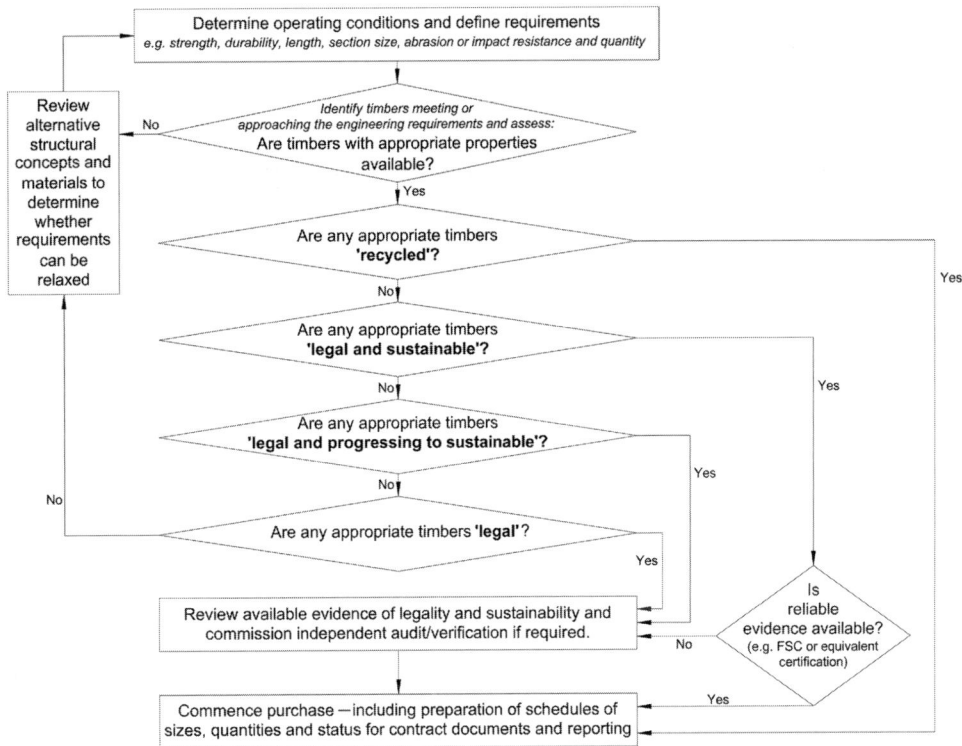

Figure 5.2. Proposed procurement framework

Table 5.2.   Definition (after ERM, 2002) and example evidence of legality and sustainability

Status	Definition	Example of evidence
Recycled	Wood which has previously been used in a structure, with or without further processing	Visual inspection (if appropriate) to ensure virgin timber is not supplied
Legal and sustainable	Legal (as below) and management of the forest conforms to international protocols relevant to sustainable forest management	As legal (below) and with verification of sustainable forest management practices. Likely to be certified in accordance with appropriate accredited scheme (e.g. Forestry Stewardship Council) with certificate number(s) on invoice and/or delivery note.

Table 5.2.   Definition (after ERM, 2002) and example evidence of legality and sustainability (continued)

Status	Definition	Example of evidence
Legal and progressing to sustainable	Legal (as below) and the producer is committed to, and making progress towards, legal and sustainable (as above)	As legal (below) – one might expect to see evidence (from forest management level) of sustainable forest management policies with programme demonstrating progress towards full implementation.
Legal	The producer has rights to use the forest and complies with relevant laws and codes of practice of the country in which the forest is situated. Species is not CITES listed (cf. Box 5.2).	Evidence of payment of appropriate taxes (government permits) and traceability of timber (concessionaire or logging company).

The various stages of the procurement framework are described in the following sections.

5.4.   TECHNICAL REQUIREMENTS

Technical requirements drive the whole timber selection and procurement process and are largely based on a comparison of loadings and exposure conditions (described in Chapter 6) with the properties of available timbers (see Chapter 4). As with the whole design process, useful lessons may be learnt from the performance of similar schemes, but it is important that lesser known species are not excluded only because a long term track record cannot be demonstrated. The tables in Chapter 4 provide examples of timbers with similar properties and the methodology described in Box 4.6 enables an indication of the resistance of lesser known species to marine borers to be rapidly assessed.

Whilst the principle technical requirements will relate to strength, durability and section size it is important that other technical properties, specific to the intended application, are also considered during the early stages of scheme development. Examples of such attributes include:

- Ease with which the timber may be machined: the tolerances provided by normal sawn timber are seldom adequate to ensure a completely flat surface when timber planking is used for a deck – it is therefore usual for fine sawn or 'planed all round' timbers to be used, but machining some timbers can prove very difficult (and expensive).

- Propensity to splinter – many tropical hardwoods will cause severe swelling if splinters enter the skin – could be ameliorated by selecting timbers less likely to cause such a reaction and/or those with less propensity to splinter.

- Ability to withstand shock loads – some timbers have a reputation for withstanding large, short duration loads (such as vessel forces on fenders) better than others – which may not be reflected in the strength properties.

5.5. AVAILABILITY OF TIMBER

The availability of timbers at the time of the scheme will also have a significant impact on the selection process. One of the first activities should be an assessment of the availability of reclaimed or recycled timber from structures due to be dismantled or local stockpiles. The practicalities of assessing timber in existing structures and accurately predicting the proportion that may be reclaimed will often be such that it is only possible to make approximate estimates and ensure that any excess material is stored ready for reuse on another occasion.

The next stage should be discussions with timber merchants regarding the availability of appropriate materials. The procurement process will often be most satisfactory if the specification is kept as simple as possible and timber suppliers encouraged to propose alternative or lesser known species for comparison with more traditional timbers. Care should be exercised when considering the use of different timber species within a single structure. For example, whilst a lock gate might use a hardwood frame with softwood planking, the relatively small volume of timber used is unlikely to present a problem to the producer or supplier. Conversely, the use of different species for the piles and planking of a large groyne field may cause considerable problems if normal processing involves the production of planks from the outer parts of the same logs that are processed for the piles.

The availability of timber (particularly certified hardwoods and non-standard section sizes or lengths) is highly dependent on the timescale for procurement. The felling and processing of timber can be a seasonal activity and additional time may be required for seasoning and transporting the timber as well as chain of custody or environmental auditing or certification activities. Whilst a number of timber merchants have a stock of traditional tropical hardwoods available in a range of dimensions, there will be considerable benefits in commencing the procurement process as early as possible. This may result in the availability of custom dimensions, particularly long lengths and large volumes, but is likely to necessitate the separation of timber supply from any construction contract.

The availability of timber will also be related to cost and whilst this is not included within the procurement framework, it cannot be wholly excluded from procurement considerations in most situations. High quality timber traditionally used in joinery is unlikely to be appropriate for use in coastal and river engineering.

5.6. RECYCLED TIMBER

The procurement framework (see Figure 5.2) clearly indicates that the use of recycled timber is to be preferred over all other appropriate timber. This has benefits in so far as it reduces the demand for virgin timber and reduces the volumes of material going to landfill. Where timber from old structures being dismantled can be immediately reused in another structure on the same site (such as the use of planks cut from an old groyne as plank piles for a new one) there are significant environmental and economic benefits in reducing the requirements for duplicated handling and transportation. Where there is no immediate use for reclaimed materials (or they are not suitable for reuse in engineering structures) they may either be stockpiled or offered to another organisation or timber dealer. A list of companies and other organisations active in recycling timber within the UK is available from SALVO, 2003.

It is important to ensure that any timber purchased as recycled is fit for purpose, as the use of substandard materials may actually be damaging if parts of the structure are placed under a greater load due to the failure of elements. It is important therefore that the materials are inspected (by qualified strength graders if appropriate) to ensure that they are suitable. In some cases recycled timber may command a price premium and in this case it may also be appropriate to verify the source of the recycled materials, particularly if they have been reprocessed, to ensure that virgin timbers are not being presented as recycled.

5.7. CERTIFICATION AND VERIFICATION

Certification enables the legality and sustainability of timber to be readily verified. Whilst it is not a prerequisite to timber being legal and sustainable, most producers will have timber certified if they can since the costs are offset by the premium and wider market that certified timber attracts. However, it must be understood that not all certification schemes are comparable, and certification systems are not immune to abuse by unscrupulous individuals or organisations.

5.7.1. Certification – proving sustainable forestry

The word certify comes from the Latin, *certis ficare*. *Certis* means 'sure', *ficare* means 'to make'. The dictionary definition reflects these Latin terms – 'to make certain, to attest in an authoritative manner'. The certification process for timber is shown in Figure 5.3 and has to be applied separately to both of the following:

1.   **Forest management certification** – Forest management certification: *"is a process that results in a written quality statement (a certificate) attesting origin and production status of a raw wood material following validation by an independent third party"* (Baharuddin and Simula, 1996). It means that

standards (environmental, legal and social) are observed to be met during its production.

2. **'Chain of custody' certification** – Chain of custody certification is third-party verification that there are adequate systems for tracking certified wood from forest source to final point-of-sale in such a way that at all times it is possible to distinguish certified from uncertified wood (see Figures 5.4 and 5.5).

In addition to these two separate certification procedures, a label will commonly be fixed to the articles for sale to the public in order that they can distinguish certified materials or products from otherwise identical but uncertified products. Labelling of a product is an optional procedure that may or may not accompany certification. It is a shorthand message for a prospective purchaser to facilitate selection or to advertise a particular brand. To be meaningful in forestry terms, labelling must be supported by certification. Construction materials are frequently not directly labelled, but the certification reference number(s) (e.g. SGS-COC-0755/65250184) should be quoted on invoices and/or delivery notes. These can be verified directly with the accreditation organisation (such as the Forestry Stewardship Council) often through a website.

Figure 5.3. Certification process

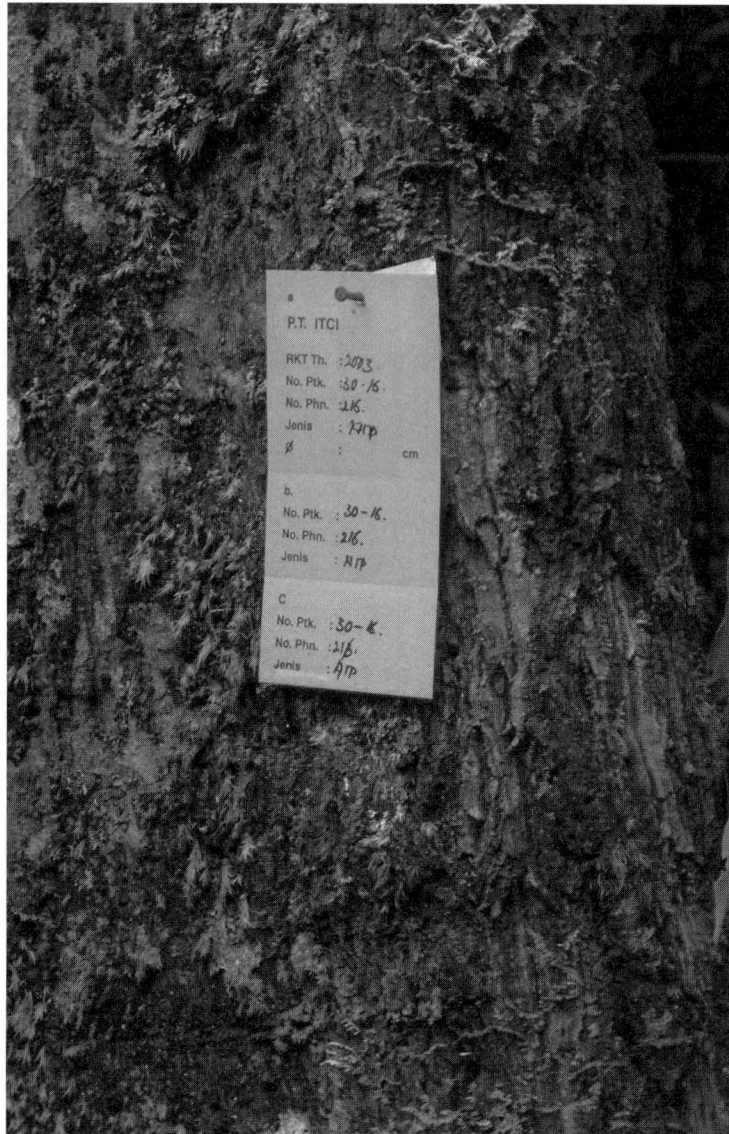

Figure 5.4. Marking the standing tree in the forest (courtesy Ita Rugge, Timber Trades Federation)

The certification process is illustrated in Figure 5.3.

Figure 5.5. Labelling the log (courtesy Ita Rugge, Timber Trades Federation)

5.7.2. Forestry standards

An appropriate forest management policy must be in place to achieve forest management certification. This comprises a *"set of definitions of minimum acceptable levels of forest management or practice"* (European Forestry Institute, 2000). A standard is a document structured upon a hierarchical concept of how an activity or a process might be performed. Thus activities or processes take place within the framework of principles – supported by criteria which, in turn, are supported by verifiable indicators of performance.

Principles: fundamental laws or rules, serving as the basis for reason or action

Criteria: defining characteristics or practices by which forest planning or forestry activity can be assessed

Indicators: Variables, attributes or components that can be measured in relation to a specific criterion

Verifiers: Sources of information or reference values for the indicator

Forestry standards require that definitions be agreed and compromises must be hammered out. In practice, it can be assumed that no stakeholder in forestry standard development will get all that they would ideally wish for, and a standard will be a document which defines the compromises which stakeholders are prepared to make in order to achieve stability. Forestry standards which are onerous and difficult to meet will naturally have a negative economic impact on an enterprise attempting to meet them. Conversely, the lack of standards or acceptance of very weak standards make nonsense of sustainable or even responsible forest management.

A good example of widely recognised forest standards are the principles and criteria adopted by the Forestry Stewardship Council which are shown in Box 5.5. It should be noted that there are a range of other forestry standards which are not always directly comparable and it would not be sensible for non-specialists to try to determine the acceptability or otherwise of particular standards for a given application. At the time of writing it was proposed (ERM, 2002) that within the UK a central resource would review the most widely established standards and provide written guidance on them.

5.7.3. Accreditation

Accreditation is often confused with certification. Accreditation can be described as a 'quality control over membership' or 'control and harmonisation of standards of certification practice'. The purpose of accreditation is to allow fair competition in a global market for certification services and it should be noted that certification bodies do not necessarily operate an exclusive arrangement with a particular accrediting organisation. The two organisations require separate and quite different structures and the relationship between them is illustrated in Figure 5.6.

Figure 5.6. Accreditation and certification bodies

5.7.4. Comparing certification systems

Whilst there are a number of potential systems of certification with possible relevance to purchasers of marine timber, evaluation of these is a difficult topic which is beyond the scope of this manual. However, the authors believe that in order to be effective in promoting sustainable forest management, a certification system must strike a balance between social, environmental and economic interests in forests. 'Industry only' certification systems, whilst they might have good technical expertise, are not sufficiently independent of commercial interest to ensure sustainability objectives.

The Forestry Stewardship Council (FSC) is one of the largest and most widely respected accreditation organisations. As a pioneer of forest certification, it tends to be the yardstick by which all other certification systems are assessed and is the only system with broad support from international environmental organisations such as Greenpeace, WWF and Friends of the Earth. It also has a well developed labelling system and a real market profile. The organisational structure is shown in Box 5.4 and the principles and criteria (after FSC, 2003) are summarised in Box 5.5.

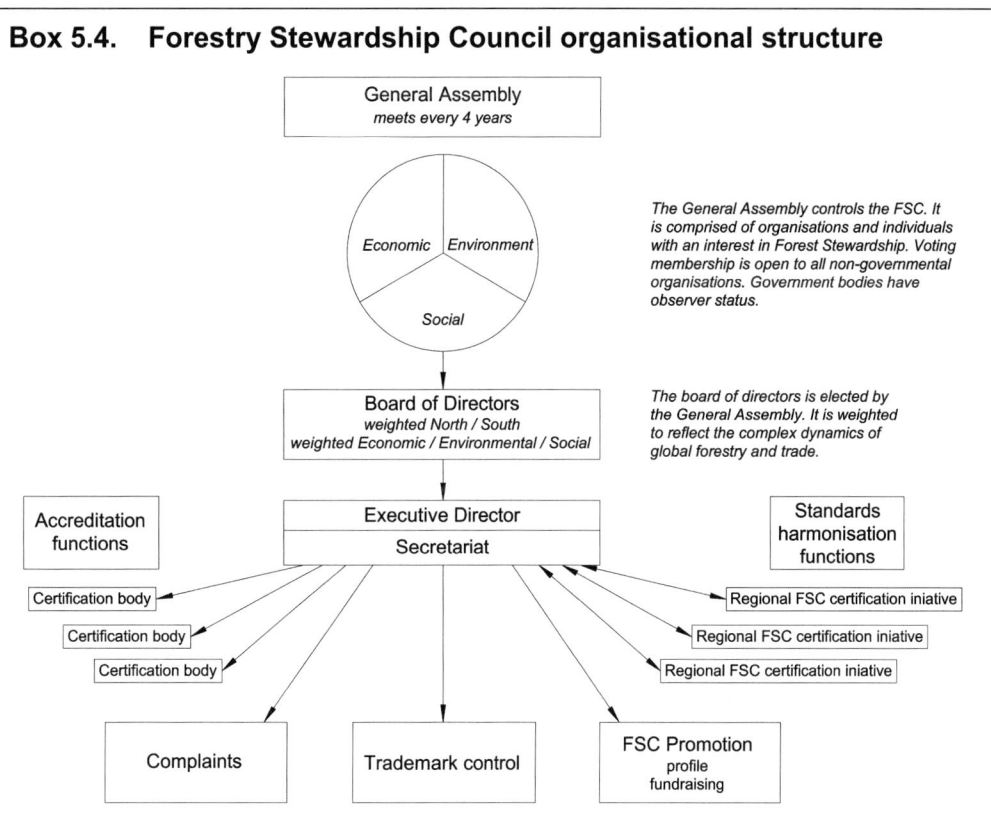

Box 5.4.   Forestry Stewardship Council organisational structure

General Assembly
meets every 4 years

Economic *Environment*

Social

The General Assembly controls the FSC. It is comprised of organisations and individuals with an interest in Forest Stewardship. Voting membership is open to all non-governmental organisations. Government bodies have observer status.

Board of Directors
weighted North / South
weighted Economic / Environmental / Social

The board of directors is elected by the General Assembly. It is weighted to reflect the complex dynamics of global forestry and trade.

Accreditation functions

Executive Director

Secretariat

Standards harmonisation functions

Certification body

Certification body

Certification body

Regional FSC certification iniative

Regional FSC certification iniative

Regional FSC certification iniative

Complaints

Trademark control

FSC Promotion
profile
fundraising

> **Box 5.5. Forestry Stewardship Council principles and criteria**
>
> 1. Compliance with laws and FSC principles
> Forest management shall respect all applicable laws of the country in which they occur, and international treaties and agreements to which the country is a signatory, and comply with all FSC principles and criteria.
>
> 2. Tenure and use rights and responsibilities
> Long term tenure and use rights to the land and forest resources shall be clearly defined, documented and legally established.
>
> 3. Indigenous peoples' rights
> The legal and customary rights of indigenous peoples to own, use and manage their lands, territories and resources shall be recognised and respected.
>
> 4. Community relations and workers' rights
> Forest management operations shall maintain or enhance the long term social and economic well-being of forest workers and local communities.
>
> 5. Benefits from the forest
> Forest management operations shall encourage the efficient use of the forest's multiple products and services to ensure economic viability and a wide range of environmental and social benefits.
>
> 6. Environmental impact
> Forest management shall conserve biological diversity and its associated values, water resources, soils, and unique and fragile ecosystems and landscapes, and, by so doing, maintain the ecological functions and integrity of the forest.
>
> 7. Management plan
> Appropriate to the scale and intensity of the operations. It shall be written, implemented and kept up-to-date. The long term objectives of management, and the means of achieving them, shall be clearly stated.
>
> 8. Monitoring and assessment
> Monitoring shall be conducted – appropriate to the scale and intensity of forest management – to assess the condition of the forest, yields of forest products, chain of custody, management activities and their social and environmental impacts.
>
> 9. Maintenance of natural forests
> Primary forests, well developed secondary forests and sites of major environmental, social or cultural significance shall be conserved. Such areas shall not be replaced by tree plantations or other land uses.
>
> 10. Plantations
> These shall be planned and managed in accordance with Principles 1-9. Plantations should complement the management of, reduce pressures on, and promote the restoration and conservation of natural forests.

5.8. SPECIFICATION, PURCHASE AND REPORTING

Standard specification clauses for both the materials and construction of timber structures are provided in Appendix 2. It is good practice to specify the required properties or range of suitable species if at all possible, since it can help to reduce environmental impacts, costs and procurement timescales.

The two main methods of procurement of timber for construction are described below.

1. Ex stock: Timber is purchased directly from stock held locally by the timber supplier. This enables the materials to be dispatched quickly and inspection is possible prior to dispatch. However, costs are likely to be higher due to storage and double handling and the range of section sizes and lengths will be limited which may lead to significant wastage.

2. Forward shipment: Timber is purchased (usually through an agent or timber supplier) from the producing sawmill and delivered directly to site. This has the advantage that the specific section sizes and lengths required can be produced by the sawmill, but the timescale is significantly longer (usually three to six months) and there can be problems with shipping, production and quality which are outside the control of the local agent or timber supplier. The cost is often denominated in a foreign currency which can lead to significant risks if the timber is purchased by the contractor as part of a lump sum contract or budgeted for some time in advance of procurement.

It is important that procedures for inspection of materials on delivery to site, verification and reporting of the actual quantity and legal/environmental status of the material are established in advance of any order. It is strongly recommended that the legality and sustainability of appropriate timbers be discussed with potential suppliers at an early stage in a project (or, for large organisations with regular projects using timber, a continuous dialogue may be maintained). It is recommended that appropriate records are kept to enable specifiers, contractors and clients to explain clearly the steps taken to ensure responsible procurement of timber and report the volume and status of timber used. A sample schedule for monitoring and reporting of timber procurement is presented in Table 5.3.

Table 5.3. Procurement schedule (after Certified Forest Products Council, 2002)

Specification reference	Product description	Status *	Submittal required	Submittal confirms certification	Supplier	Chain of custody number	Evidence attached ?
			✓ or ✗	✓ or ✗		____ - ____ - ____	✓ or ✗
			✓ or ✗	✓ or ✗		____ - ____ - ____	✓ or ✗
			✓ or ✗	✓ or ✗		____ - ____ - ____	✓ or ✗
			✓ or ✗	✓ or ✗		____ - ____ - ____	✓ or ✗
			✓ or ✗	✓ or ✗		____ - ____ - ____	✓ or ✗
			✓ or ✗	✓ or ✗		____ - ____ - ____	✓ or ✗
			✓ or ✗	✓ or ✗		____ - ____ - ____	✓ or ✗
			✓ or ✗	✓ or ✗		____ - ____ - ____	✓ or ✗

*(legal and sustainable / legal and progressing to sustainable / legal)

6. Design process

Timber has a natural resilience which enables it to withstand shock loads such as those caused by wind, waves or vessels better than most other structural materials and this resilience has considerable advantages in coastal and river situations where impact loads often produce the highest stresses. However, to achieve the greatest benefit from what is a finite and valuable resource it is important that the design process results in structures that are both efficient and effective.

In the context of the structure life cycle (see Figure 2.2) this chapter deals with the detailed design stage which normally follows scheme development and data collection but precedes procurement of materials and construction activities. The design process basically involves collecting input data, analysis and producing output data. The analysis can be broken down into the seven stages shown in Figure 6.1 and as the figure indicates, the analysis is often an iterative process. As each stage is undertaken new information may be derived which needs to be fed back for the earlier analysis to be repeated as necessary.

At the end of the process, a design should be achieved which conforms with all the input criteria. It should also give an acceptable estimate of whole life costs taking account of both capital, operation, maintenance and renewal costs (see Section 2.3.1). Maintenance and renewal cost schedules should reflect:

- the need to renew damaged, abraded or decayed members
- the need to replace or re-secure fixings (bolts, coach screws, etc.)
- costs of mobilising labour and equipment to site
- the possibility of prefabricating some elements in such a way as will facilitate future replacement.

Where possible, the design process should be informed by a review of the nature, performance and condition of similar structures in the locality operating under comparable conditions (Box 6.1) together with any monitoring records.

Regular design reviews and design checks are also essential and the findings should be treated as new input data. The output data will vary depending on the arrangements for implementation, but will generally comprise working drawings and supporting specifications together with a schedule detailing the exact numbers and dimensions of the timber components which is often incorporated into a bill of quantities.

MANUAL ON THE USE OF TIMBER IN COASTAL AND RIVER ENGINEERING

Input data	Analysis	Output
• Option appraisal	Review choice of timber structure	
• Consents		
• Operating conditions	Determine operating conditions and functional requirements	
• Design life		
• Material properties	Select suitable timber	• Whole life costs estimate
• Design codes		• Drawings
• Existing structures	Design overall configuration of structure	• Specifications
• Proprietary information		• Schedule of components
• New information analysis	Design overall stability of structure	• (Bill of quantities)
• Design reviews	Design members and connections	
• Design checks		
• Maintenance requirements	Design finishes and fittings	

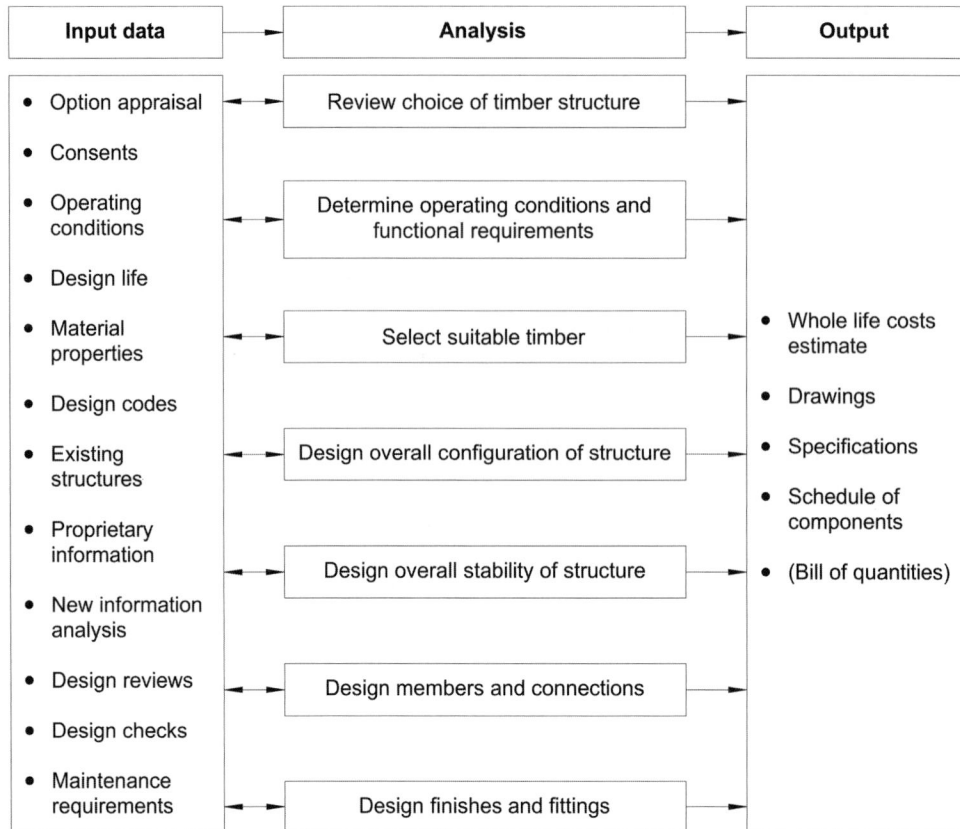

Figure 6.1. Design methodology

Box 6.1. Examination of existing structures

In designing new structures it is good practice to learn as much as possible from comparable existing structures operating under similar conditions. Examinations can be undertaken concentrating on the following aspects of the nature, condition and performance of such structures.

Construction details

- timber species
- overall configuration
- members and connections
- finishes and fittings

Defects

- overall instability (settlement, sliding, rotation)
- failed members (sheared, split, crushed)
- failed connections (disconnected, loose, displaced)
- failed fittings (disconnected, loose, displaced)

> **Box 6.1.   Examination of existing structures** (continued)
>
> Wear and tear
>
> - due to water action (location, pattern, severity)
> - due to sediment action (location, pattern, severity)
> - due to operational impacts (location, pattern, severity)
>
> Local impacts
>
> - on human environment (users, commerce, amenity)
> - on built environment (other structures, infrastructure)
> - on natural environment (processes, land forms, habitats)
>
> Estimate of residual life of structure
>
> - taking account of date of construction, present condition and future anticipated loading and deterioration

This and the following chapter provide guidance on good practice and issues that should be considered during detailed design and construction of timber structures on open coasts and in estuaries, ports, harbours, rivers and canals.   However, it is assumed that the reader is familiar with standard civil engineering concepts and procedures, and reference should still be made to the relevant national or international standards and codes of practice and other appropriate design references (Box 6.2). There is necessarily a degree of generalisation and structure specific guidance, together with appropriate references, is given in Chapter 8.   Reference to experienced engineers and specialists may also be necessary where particular problems are envisaged or encountered.

> **Box 6.2.   Relevant codes of practice and design references**
>
> **BS 6349:** *Maritime structures.*
>
> **DD ENV 1995-1.1: 2002** *Eurocode 5: Design of timber structures. General rules and rules for buildings.*
>
> **BS 5268 Part 2: 2002:** *Structural use of timber. Code of practice for permissible stress design, materials and workmanship.*
>
> **BS EN 338: 1995:** *Structural timber. Strength classes.*
>
> **BS EN 844 Part 1: 1995:** *Round and sawn timber. Terminology. General terms.*
>
> **BS 8004: 1986:** *Code of practice for foundations.*
>
> **BS 5930: 1999:** *Code of practice for site investigation.*
>
> **BS IS 15686**: *Service life planning of buildings and constructed assets.*

Box 6.2.   Relevant codes of practice and design references (continued)

Beach Management Manual (CIRIA, 1996)

Water Practice Manuals – River Engineering (IWEM, 1987, 1989)

Coastal Engineering Manual (USACE, 2002)

Timber Designers Manual (Ozelton and Baird, 2002)

MAFF (now Defra) Flood and Coastal Defence Project Appraisal Guidance:
- FCDPAG1: Overview (including general guidance) (2001)
- FCDPAG2: Strategic Planning and Appraisal (2001)
- FCDPAG3: Economic appraisal (1999) and supplementary note March 2003
- FCDPAG4: Approaches to Risk (2000)
- FCDPAG5: Environmental Appraisal (2000)

6.1.   CHOICE OF TIMBER STRUCTURE

The first stage of the design process is to review the option appraisal that led to the choice of a timber structure. The original input data needs to be re-examined together with any additional information that has arisen and then reapplied. In some circumstances it may be necessary to undertake further investigations, especially if there are significant gaps in the input data set. In terms of whole life costs this first stage is usually the most important stage of the design process.

6.2.   OPERATING CONDITIONS AND FUNCTIONAL REQUIREMENTS

The loads and conditions from the natural and human environment are described in the following section. When determining appropriate design parameters it is important to consider the proposed life span of the structure, its desired performance, the likely maintenance regime and the consequences of failure. In practice it is rarely possible to determine all of the operating conditions accurately. However, often secondary considerations such as abrasion and scour can be just as important as the primary considerations of, say, wave and soil loadings. Design conditions should also take into account situations that might only arise during the construction stage (e.g. loading from construction plant).

Much of the information required to determine loadings for use in the design of maritime structures can be found in Maritime Structures Code – BS 6349 (British Standards, 2000) or other design references such as the Beach Management Manual (CIRIA, 1996), Coastal Engineering Manual (USACE, 2002) and Water Practice Manuals 7 and 8 on River Engineering (IWEM, 1987, 1989). Other relevant loadings may be assessed with reference to BS 6399 (Loading for buildings) or proprietary design manuals.

In assessing loadings it is important that in addition to the individual loads, the combined effects of different loads acting at the same time are considered. In the case of coastal structures this frequently involves an assessment of the joint probability of waves and water levels which is described in the Beach Management Manual (CIRIA, 1996). The effects of climate change and variability should also be considered and in the UK guidance is available from the Environment Agency (2003).

The information provided in the sections below covers most of the more common considerations in respect of timber structures used in coastal and river engineering. Due to their long term nature, environmental loads such as the effects of snow, ice, temperature and time averaged wind are not considered as dynamic loads.

6.2.1. Ground conditions

Assessment of ground conditions for foundations is described in BS 8004: 1986 and site investigations in BS 5930: 1999 both of which are relevant to coastal and river areas. In these situations competent substrates are often overlain by one or more layers of soft and/or mobile sediments. The characteristics and extent of such mobile sediment may fluctuate in level and the introduction of a new structure can frequently have a significant impact upon this. It is important therefore that careful consideration is given to the extent to which soft and mobile sediments are expected to contribute to the overall stability of any structure. Particular attention will need to be given to the stability of artificially raised or lowered ground and/or bed levels in the presence of wave action or tidal/river flows.

The combination of certain ground conditions (particularly sand or gravel beaches) in combination with environmental loads can lead to significant abrasion of structures. Conversely, other ground conditions (silt or mud) may contribute to the preservation of parts of timber elements by preventing oxygen reaching them.

6.2.2. Biological attack and water conditions

The temperature and degree of salinity, contamination or turbidity of the water in which timber structures are located can have impacts on the durability of the structure and particularly any metal fixtures and fittings. However, this is often less significant than biological attack which is at least partially correlated to the water conditions and may vary dramatically with time. The degradation process is discussed in Section 9.1.

Biological attack in fresh water

Fungal decay (see Section 4.5.2) is the most important biological agent that may attack timber in fresh water. Such decay is most prevalent in timber with a moisture content above 20% – this is likely to occur in timber near and above the water line as well as that embedded in concrete, brick or other porous materials and poorly ventilated, horizontal surfaces. Aggressive fungal decay above the water level may occur, principally by wet rot type fungi. In addition, insects may also attack wet timber above water level, although insect attack may be symptomatic of fungal decay being present in the structure.

It is worth noting, however, that timber that is permanently and completely immersed in fresh water will normally only suffer fungal decay of the soft rot type. Soft rotting type fungi erode the outer layers of the timber components at a relatively slow rate. The outer surfaces are typically dark, soft and 'cheesy' in texture. The depth and rate of penetration of the fungi is largely governed by density. However, the affected timber may also be exposed to fast currents and abrasive conditions. The combined effects of the soft rot fungi and faster erosion of the decayed outer layers of timber may accelerate the deterioration of the component. Pockets of fungal decay detected below the waterline are usually attributable to decay migrating down through the timber component. This observation is also common in marine structures that have been weakened by pockets of wet rot type decay.

To prevent biodeterioration, timbers in structural positions in which decay might be expected (given the foregoing advice) should be naturally durable, or preservative treated, legislation permitting.

Biological attack in the marine environment

Salt water can act as a timber preservative and heavy salt deposition in wood may offer some protection against fungal decay for timber out of contact with sea water although it is important to realise that salt deposition may not be permanent. Horizontal faces of timber or joints are at risk of fungal and insect attack if the moisture content of the components rises above the decay threshold of 20% for prolonged periods. Wood that is located below the high tide level may be subject to attack by marine borers, bacteria and fungi.

The principal agents causing biodeterioration of timber in the marine environment are marine borers, which can cause severe damage in relatively short periods of time. In UK and other temperate waters two types of borer are of greatest significance; the mollusc *Teredo* spp. (shipworm – see Box 4.5) and the crustacean *Limnoria* spp. (gribble – see Box 4.4). The former has commanded particular attention owing to the inherent difficulties in accurately estimating the extent of its attack.

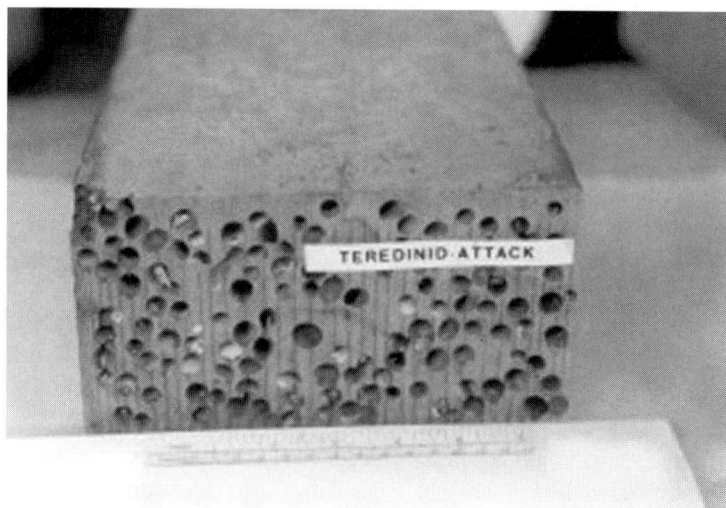

Figure 6.2. Teredinid attack (courtesy Simon Cragg)

Although the incidence of marine borers (for details see Section 4.5.2) is an important consideration for the use of timber structures in coastal and maritime engineering, knowledge of the severity of attack around the UK, and indeed European, coastline is limited. It is thought that geographical features such as changing water temperatures heavily influence overall distribution of the organisms, and other factors within harbours and river estuaries such as tidal range, sea and air temperatures, salinity, oxygen levels and pollutants may also be significant. The last comprehensive survey of UK waters was undertaken during the 1960s, and indicated that *Limnoria* spp. is capable of existing in one species form or another around much the UK coastline, but *Teredo* spp. tends to be limited to the South West coastline although it has been reported in more northerly waters. There is concern that with increasing sea temperatures and less polluted estuaries, the extent of these organisms may have increased and the European standard which classifies exposures into a series of hazard classes (BS EN 335 Part 2: 1992) suggests that marine borers are universally present within Europe. The classification of European hazard classes is presented in greater detail in Section 6.3.1.

It should be noted that the occurrence of marine borers is sporadic and the severity of attack often varies from site to site and season to season. A definitive understanding of the hazards of marine borers thus requires intensive sampling over a number of years to minimise errors due to seasonal and annual fluctuation. One factor thought to be of importance is the influence of marine borer infested wrecks and other redundant timber structures as well as driftwood in and around harbours. The risks of introducing borers through wrecks and driftwood may be minimised by good harbour hygiene. This may not prove to be as important a factor where beach/flood defences are erected as driftwood may be carried away by the prevailing tides. Various approaches to reduce the impacts of biological attack are summarised in Table 6.1.

Table 6.1.   Overview of methods of mitigating marine borer attack (after DWW, 1994)

Method	Advantages	Disadvantages
Preservative treatment Applying preservatives under pressure	• Protection against marine borers • Permeable timber species are easy to treat	• Potential leaching of preservatives into environment • Timber has to be treated before construction
External treatment Applying a protective coating over the timber	• Protection against marine borer attack	• Coating may be damaged (regular maintenance required) • Timber has to be treated before construction • Protective coat may damage the environment

Table 6.1. Overview of methods of mitigating marine borer attack (after DWW, 1994) continued

Method	Advantages	Disadvantages
Sealing Wrapping the timber in an impermeable covering	• Protection against marine borers • Kills marine borers already present in the structure • Can be applied after construction	• Expensive • Covering may be damaged • Does not increase the structural capacity
Detailing and construction Reduce number of components exposed and/or use durable timbers	• Minimal environmental impact	• Restricted construction possibilities • May require larger section sizes
Use of alternative materials	• Reduced susceptibility to marine borer attack	• Inappropriate use of non-renewable materials • Often lower ecological value

6.2.3. Water levels

Water levels are often subject to variation at a number of different timescales. Rivers change seasonally and levels may be subject to relatively short duration peaks during high flow or flood events. Sea levels are often dominated by tidal cycles, but may also be affected by storm surges and wind or wave set-up. Short period waves including those generated by the local wind field, swell waves propagating from further away and boat waves, have a much higher frequency and can often be superimposed on the still water levels (except of course where wave set-up is significant).

Changing water levels can impact on the design of structures in a number of ways:

- the need to meet performance requirements over a wide range of conditions
- the need to cope with hydrostatic or hydrodynamic pressures due to different water levels resulting from wave action, tidal lag, etc.
- the need to cope with variable ambient conditions ranging from constant saturation, to cycles of wetting and drying, to intermittent spray and precipitation.

Hydrostatic loading has to be considered when there is water pressure on one side of an immersed structure but no equivalent resisting pressure on the other. The most common example of this situation in the coastal and river environment is the lock or river control gate where a higher water level is retained on one side of the gate. Timber plank piling is sometimes used for river bank protection and can also be subject to hydrostatic loading when there is a rapid draw-down of river water levels leaving the water levels temporarily higher behind the piling. Groynes can also be affected when a wave crest is located on one side of the structure and a trough on the other.

Pressure intensity within a liquid varies with depth. It also varies in proportion to the density of the liquid. The net force on an immersed plane surface is the product of the area and the hydrostatic pressure intensity at the centroid of the surface. To calculate this force:

Force $F = A.\rho.g.h$

where   A is the area of the surface   g is gravity
              ρ is the density of the liquid   h is the depth of the centroid of the surface

The net force acts at the centre of pressure at a depth H below the surface and:

$$\text{Depth } H = h + \frac{I_{centroid}}{A \cdot h} \quad I_{centroid} \text{ is the moment of inertia of the area}$$

For a rectangular plane area with one edge in the surface the net force F acts at a depth of two thirds of the total depth of the rectangle.

6.2.4. Current action

Almost all structures constructed in a coastal or river situation will be subject to currents. For permanent structures the maximum current speed expected at the site for a return period similar to the anticipated scheme life is normally used for design. Currents are usually more significant in river locations as compared to coastal situations where waves frequently dominate, however currents cannot be discounted at some coastal structures, such as impermeable groynes, where they may cause scour at the end of the groyne or wear on particular parts of the structure.

Loads imposed directly by tidal or river currents on structures can be classified as:

- drag or in-line forces, parallel to the flow direction
- cross-flow forces, transverse to the flow direction.

Current drag forces are principally steady, any oscillatory components only becoming significant when their frequency approaches a natural frequency of the structure.

Cross-flow forces are entirely oscillatory for bodies presented symmetrically to the flow.

The steady drag force for a uniform prismatic structural member immersed in a uniform current can be calculated from:

$$F_D = {}^1\!/_2\,(C_D.\,\rho.\,V^2.\,A_n)$$

where F_D is the steady drag force in kN
         V is incident current velocity in m/s
         C_D is the drag force coefficient
         A_n is the area normal to the flow in m^2
         ρ is the density of water in tonnes/m^3

The force acts at the centroid of the area normal to the flow.

Values of C_D and A_n should be determined taking due account of the effect of marine growth. Values of current drag force coefficients for piles, both circular and non circular, can be found in BS 6349-1: 2000.

Where waves and currents combine to increase drag forces, then water particle velocities should be added vectorially and the result used to calculate the drag force.

Circular sections such as piles situated in a current, experience both in-line and cross-flow fluctuating forces due to vortex shedding downstream of the section. The frequencies of these forces is directly related to the frequency of the vortex shedding. The amplitude of the fluctuating force increases as its frequency approaches the natural frequency of the section, or that of the whole structure. Piled structures are particularly vulnerable to this type of oscillation during construction and pile heads may need to be restrained or damped to prevent it. The critical-flow velocity V_{Crit} is given by:

$$V_{Crit} = k.f_N.W_s.$$

where f_N is the natural frequency of the section
         W_s is the diameter of the section
         k is a constant equal to:
             1.2 for onset of in-line motion
             2.0 for maximum amplitude of in-line motion
             3.5 for onset of cross-flow motion
             5.5 for maximum amplitude of cross-flow motion

BS 6349-1: 2000 provides more information on the use of this equation.

6.2.5. Wave action

The loads imposed by wave attack have a major influence on the design and construction methods used for the building of structures in the coastal and river environments. A knowledge of the wave activity and persistence in both the average and extreme conditions at the chosen site is an essential prerequisite for the design.

Where the maximum stresses in a member of a structure due to wave loading consist of more than 40% of the maximum total combined stresses then it will be necessary to check the fatigue life of that element.

The design wave parameters required in an ultimate load analysis of structures having quasi-static response characteristics to wave loading can be taken as the height and period of the average maximum incident wave having a return period of 50 years. A quasi-static analysis assumes the displacement is equal to the loading multiplied by an impact factor and divided by the static stiffness of the structure.

Where there is a significant dynamic response of the structure to wave loading, consideration should be given to the possible range of wave periods and associated maximum wave heights that would result in the greatest dynamic magnification. Where frequency, f_c, of a forcing cyclic load approaches the natural frequency, f_N, of the structure, the response of the structure to the forcing load is magnified relative to that predicted by quasi-static analysis. Dynamic effects are not usually significant where f_c is less than $f_N/3$ or greater than $2f_N$; f_N should be considered both for the structure as a whole and for each important element.

Design wave forces can be derived from design wave parameters either by calculation or by physical model tests.

Wave height, period, and the dimensions of the structure all contribute to the magnitude of the wave force, however the resulting hydrodynamic regime also plays a role. The relationship between the width or diameter of the submerged part of the structure or element, W_s, and the wave length, L, determine the applicable hydrodynamic regime as follows:

$W_s/L > 1$ reflection applies

$0.2 < W_s/L < 1$ diffraction theory applies (limited application to coastal/river structures)

$W_s/L < 0.2$ Morisons equation applies

BS 6349-1: 2000 provides details of the design wave parameters and the derivation of wave loads for the reflective condition, and using Morisons equation. It should be remembered that reflective structures (such as impermeable groynes), will not dissipate a large proportion of the incident wave energy and will attract much higher forces than a porous or discrete structure. Wave energy which is reflected from such a structure may contribute to increased sediment transport or scour which may, in turn, lead to increased wave exposure. Wave forces may also be large if incident waves are funnelled into a confined area.

The orientation of the structure to the incoming wave front also affects the nature of the forces on the structure. For example, a wave front approaching the shoreline at an angle does not exert a steady, sustained load over the full length of a groyne but produces a discrete load which travels along the length of the groyne as the wave impacts.

6.2.6. Wind and spray

Timber structures in the coastal and river environment are usually low and the direct effect of wind loading when compared with loads from other sources is probably small. BS 6349-6: 1989, the *Code of practice for maritime structures – Part 6: the design of inshore moorings and floating structures* provides supplementary information for CP 3: 1972: Chapter V-2, Wind loads, which is of particular relevance to vessels and floating structures in coastal waters. It also includes sufficient information to allow loads on simple structures in open waters to be calculated without reference to CP 3: Chapter V-2.

Where the timber structures are situated in open waters the CP 3: Chapter V-2 method, incorporating the BS 6349-6: 1989 modifications, could be followed which will give slightly higher wind speeds and higher loads. In more sheltered situations the standard CP 3: Chapter V-2 approach would be appropriate. However it should be noted that CP 3: Chapter V – Part 2 is now withdrawn and BS 6399-2: 1997 *Loading for buildings, Part 2: Code of practice for wind loads* is a technical revision which replaces it. The new code should be used therefore for situations that cannot be classified as 'open waters'.

The effects of wind loading may also be significant if water particles and sediment are carried by the wind, increasing the general wear and tear on a structure, including the vulnerability to corrosion of fixings and fittings.

6.2.7. Mechanical damage

Mechanical damage can take many forms including single loads from large impacts (for example, vessels hitting piers) and abrasion from multiple smaller impacts (such as the effect of beach material passing across a groyne). Traditionally, species such as elm, purpleheart, ekki and jarrah have been used for fendering because of their ability to withstand shock loads, which is thought to be related to the interlocked nature of the grain. It should be noted, however, that elm is non-durable and vulnerable to biological attack so the benefits have to be considered against the risk of deterioration. Abrasion can be an important factor in determining the service life of timber structures and is a feature of both sand and shingle beaches. Resistance of timber to abrasion is thought to be related to density, with more dense timbers (such as tropical hardwoods) generally being more resistant.

6.2.8. Loads and conditions from human activity

In addition to the environmental loads, forces imposed on the structure by human activity need to be considered. Many of these forces can be quantified using information found in BS 6349 (Maritime structures), BS 6399 (Loadings for buildings) or BS 648 (Weights of building materials).

Additional information can be drawn from proprietary design manuals (such as those referenced in Section 6.1) the most important of which are summarised in the sections below.

During construction

The main areas of impact upon the structure and its components during the construction stage are:

- unloading and storage

- handling and processing
- installation
- temporary applications.

The impacts upon the materials during unloading are likely to be small in comparison with the impacts experienced under working conditions, provided the operation is carried out in a proper manner. Storage should also involve no significant impacts, again provided it is undertaken in a proper manner. Timber members stored badly can suffer from end splitting and warping.

Handling and processing involves cutting, drilling and shaping timber members and moving them to their point of installation. Once again, the impacts involved should be well within the demands of the working conditions, provided everything is undertaken in accordance with good site practice. The temptation to sometimes force members into position may also result in over stressing although good practice should eliminate this.

Temporary applications involve the structure itself being used to facilitate construction. This might include working platforms for plant and equipment, and anchorage points for propping and jacking operations.

During operational stage

The impacts upon a structure during its operational stage can be classified according to whether or not they can be predicted with a reasonable degree of accuracy. Predictable loads include those relating to the function of the structure, such as pedestrian loads on a jetty or berthing loads on a dolphin and maintenance or other planned activities.

When determining unpredictable impacts, consider the possibility of increased loads which would be feasible given the nature of the structure. For example, a slipway intended for dinghies which is wide enough to accommodate heavy plant, or an access ladder on a jetty which could be used for mooring vessels. Accidents and vandalism should also be considered and will require a risk assessment to determine the seriousness of any failure resulting from potential scenarios. Where the consequences of failure are high, allowances should be made for such impacts.

6.3. SELECT SUITABLE TIMBER

The selection of suitable timber requires consideration of a wide range of factors including technical and environmental requirements, and costs as well as the availability of different timbers. Resolving the different issues often needs several iterations before final selection and specification of the timber. The timber procurement process is described in Chapter 5, but before that can commence the technical requirements must be defined. These requirements relate to the structural concept, loadings from the natural and human environment and functional/performance requirements (including service life). Some of the requirements can be dealt with in a formal manner (such as the use of hazard classes for biological attack described below) whilst others are more subjective. A review of practical experience of the performance of timbers in specific circumstances can often be helpful.

6.3.1. Hazard classes

Within Europe, hazard classes are used to categorise potential damage from biological agents and determine the required durability or preservative treatment of timbers used within a structure. As can be seen from the definitions in Table 6.2 the classes are quite broad, but BS EN 335: Part 2: *Hazard classes of wood and wood-based products against biological attack* does introduce the use of sub-classes, allowing further refinement. Most timber structures falling within the scope of this document will be within HC4 or HC5 although some components above the water level come under HC3.

Table 6.2.   Hazard classes

Hazard categories		Environment
BS EN 335	BS 5589	
HC1	1	Situation where the wood or wood-based product is under cover, fully protected from the weather and not exposed to wetting
HC2	2	Situation in which the wood or wood-based product is under cover and fully protected from the weather but where high environmental humidity can lead to occasional but not persistent wetting
HC3	3	Situation in which the wood or wood-based product is not covered and not in contact with the ground. It is either continually exposed to the weather or is protected from the weather but subjected to frequent wetting
HC4	4	Situation in which the wood or wood-based product is in contact with the ground or fresh water and is regularly exposed to wetting in fresh water environments
HC5	M	Situation where the wood or wood-based product is regularly exposed to sea water

BS EN 460: 1994 *Durability of wood and wood-based products – Natural durability of solid wood. Guide to the durability requirements for wood to be used in hazardous classes* provides guidance on appropriate classes of wood durability for the various hazard classes and suggests that for situations where there is a significant risk of attack by marine organisms only the heartwood of species classified as durable or moderately durable should be used untreated. Similar considerations also apply to termites and wood-destroying beetles. More detailed guidance (in the form of Table 6.3) is provided for wood destroying fungi.

Table 6.3.   Wood-destroying fungi – durability classes of wood species for use in hazard classes

Hazard class	Durability class				
	1	2	3	4	5
1	o	o	o	o	o
2	o	o	o	(o)	(o)
3	o	o	(o)	(o) – (x)	(o) – (x)
4	o	(o)	(x)	x	x
5	o	(x)	(x)	x	x

o	Natural durability sufficient.
(o)	Natural durability normally sufficient but in some applications preservative treatment may be appropriate.
(o) – (x)	Natural durability may be sufficient but depending on end use, preservative treatment may be necessary.
(x)	Preservative treatment is normally advisable but in some applications natural durability may be adequate.
x	Preservative treatment necessary.

However, the selection of timber should not be related solely to the hazard class: the impacts of failure, the ease with which the element or structure may be repaired or replaced and its service life should also be considered.

Preservative treatment is described in Chapter 4; other aspects of the selection and procurement of timber are described in Chapter 5. It is highly likely that technical requirements for timber may change during the design process as particular issues become more or less significant. The selection of appropriate species is likely therefore to be an iterative process, with significant refinements being made during the scheme development and detailed design period.

6.4.   OVERALL CONFIGURATION OF STRUCTURE

This deals with the overall shape, dimensions and nature (e.g. permeability) of the structure and is driven mainly by the function, operating conditions and performance requirements for the structure. Typically, the overall configuration is not dependent on the choice of construction material but rather related to the particular type of structure (e.g. groyne, dolphin, river training wall) and is therefore largely outside the scope of this manual. However, some structure specific guidance is provided within Chapter 8.

6.5.   OVERALL STABILITY OF THE STRUCTURE

This addresses the stability of the structure as a whole in terms of settlement, uplift, sliding and rotation.   At this stage in the design process it is assumed that the structure is stable within itself.   Normally, overall stability is primarily achieved by foundation arrangements.   Useful information on the design of foundations is provided in BS 8004: 1984 *Code of practice for foundations*.

For many timber structures used in coastal and river engineering, the starting point is single, large section, cantilever piles, normally driven into the substrate. These usually lend themselves to most construction conditions and provide a sound framework for developing the rest of the structure.   However, depending upon the operating conditions, it is often necessary to modify or supplement this approach to achieve the necessary measure of stability.   The most common techniques employed are detailed below.

6.5.1. Planted posts

Where the ground conditions are too hard for the piles to achieve sufficient penetration, holes or trenches can be excavated into the ground, timber posts inserted and the excavations backfilled with arisings or imported fill such as concrete, as shown in Figure 6.3.

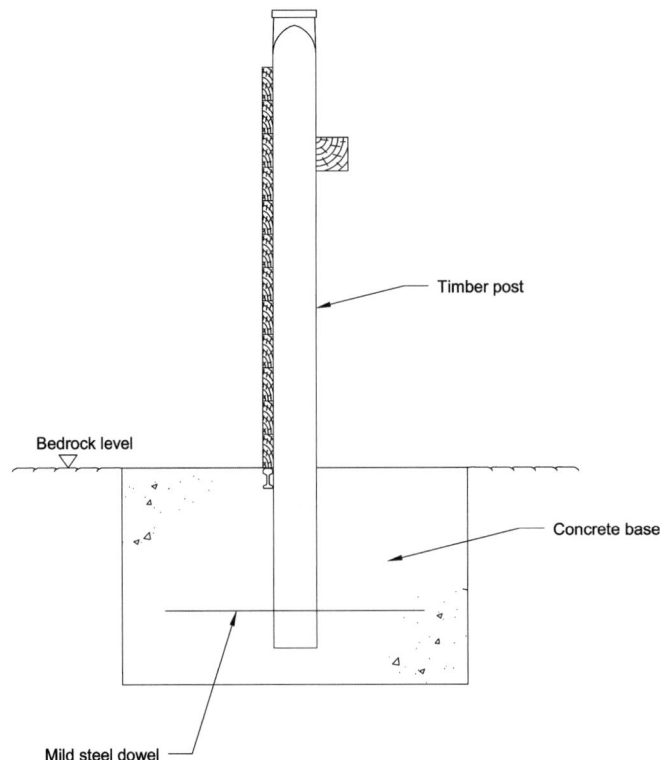

Figure 6.3. Planted post

Posts planted in concrete can also be used in situations where piling would otherwise be possible but the lengths of timber required are unavailable. Shorter lengths of timber set in a concrete base can potentially match the performance of longer driven piles.

6.5.2. Plank piles

Where the ground conditions are too soft for the piles to resist rotation, in-fill panels of smaller section piles, such as plank piles, can be installed between the main piles as shown in Figure 6.4. This approach also provides a useful method for low level work and allows the substrate to erode without necessarily compromising the structure.

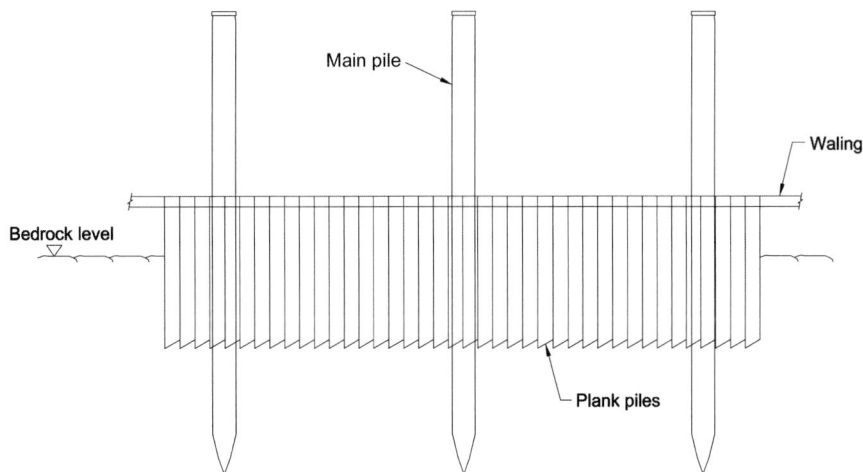

Figure 6.4. Plank piles

6.5.3. Buried panels

An alternative arrangement to the plank piles described above is to excavate between the main piles and fix horizontal in-fill panels, such as planking. In the case of groynes this would involve extending the planking below the lowest anticipated beach level, as shown in Figure 6.5. This technique has the disadvantage of involving extra work, but reduces the amount of piling involved.

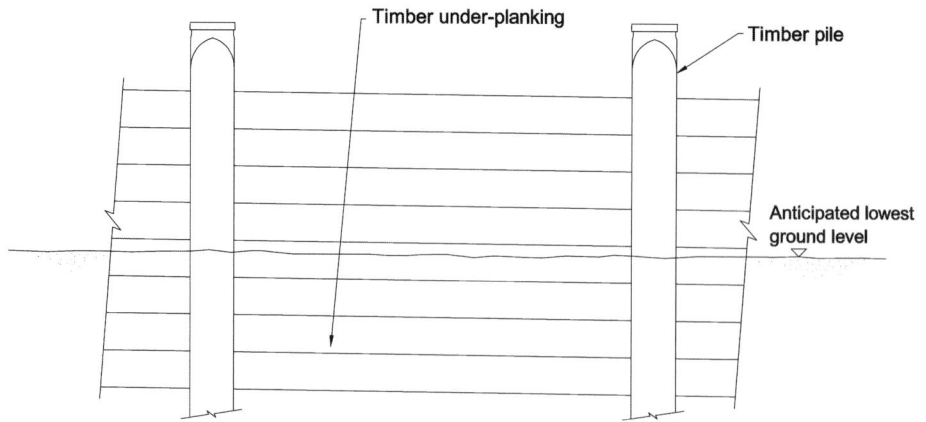

Figure 6.5. Buried panels

Buried panels are often used instead of plank piles when most of the excavation necessary is in a non-cohesive material.

6.5.4. Props

Again, where the main piles are unable to resist rotation on their own, props (normally working in compression and therefore more efficient) can be used. They will require their own anchorage arrangement (Figure 6.6). In a groyne situation a prop is located on the down-drift side, and is therefore more vulnerable to beach erosion.

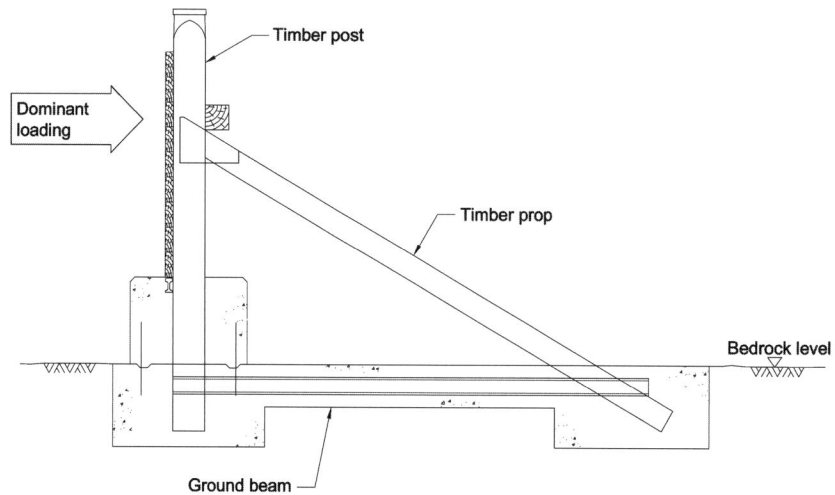

Figure 6.6. Prop arrangement

6.5.5. Ties

Once again, where the main piles on their own would rotate, ties (normally working in tension) can be used. Ties can be in timber or steel (depending upon the situation) and will require an anchorage system such as secondary piling or anchor walls (Figure 6.7). As a tie works in tension, it is structurally less efficient than a prop but, located on the up-drift side of the groyne, will be less vulnerable than a prop to beach erosion.

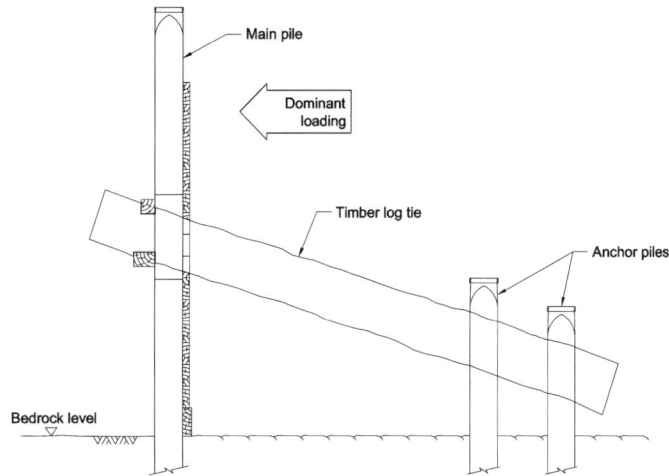

Figure 6.7. Log tie

6.5.6. Pile groups

Where the overall configuration of the structure lends itself to such an arrangement, piles can be structurally linked together to act in unison rather than individually. This can increase their effective stability considerably as shown in Figure 6.8.

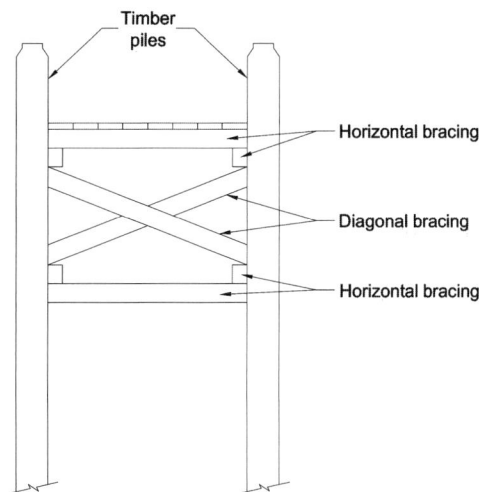

Figure 6.8. Pile group

In coastal and river engineering the loadings are often dynamic, intermittent and discrete. So, for instance, a random wave front approaching the shoreline at an angle does not exert a steady, sustained load over the full length of a groyne. Instead, at any given point in time it impacts certain lengths whilst others are left relatively unaffected. For this reason it is usually necessary to design structures with the ability to spread localised areas of high loading. This should then be taken into account when calculating the overall stability of a structure.

6.6. MEMBERS AND CONNECTIONS

A number of issues should be borne in mind when designing members and connections:

- The detailed design development of members and their connections should be undertaken in tandem. Connection details will often influence the design of the members, and vice versa.
- The number of connections should be kept to a minimum and in relation to this, individual members should be kept as long as possible.
- The arrangement of members and connections should be kept as uncomplicated as possible.
- The arrangement of members and connections should be designed to allow for the distribution of localised loads into the structure.
- Permissible stresses can generally be increased for short term loading conditions.
- Deflection is not normally a critical matter except for, say, decking for jetties and similar structures.
- If possible, some redundancy should be built into the structure.

6.6.1. Member categorisation

From a design perspective, members typically fall into 1 of 3 categories (Figure 6.9).

- Primary: for the main stability of the structure (e.g. piles, props and ties)
- Secondary: for the main distribution of loads (e.g. walings and transoms)
- Tertiary: for cladding and ancillaries (e.g. planking, decking and handrailing).

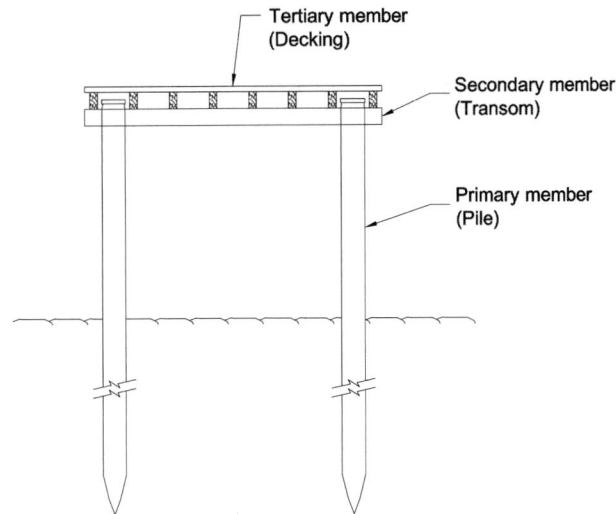

Figure 6.9. Member categories

It is usually good practice to detail the tertiary members and their connections in a manner which will maximise their ability to also distribute loads into the structure.

6.6.2. Member design

At the time of writing two alternative methods are available for designing members of timber structures. The permissible stress approach to the design of structural timber is contained in BS 5268: Part 2: 2002: *Structural use of timber. Code of practice for permissible stress design, materials and workmanship.* The limit state design philosophy is the subject of Draft for Development, Eurocode: 5 *Design of timber structures – Part 1.1: General rules and rules for buildings* (together with United Kingdom National Application Document).

Eurocode 5: Part 1.1 for timber structures was published in draft in the UK in 1994 as DD ENV 1995-1-1: 1994, together with a National Application Document for facilitating its use in the United Kingdom during the ENV (European pre-standard) period. The definitive EN EC5 is expected to be published in the UK by the end of 2004. The intention is to use the code in parallel with BS 5268 until year 2005 at which time BS 5268 would be termed 'obsolete'.

Permissible stress method

The permissible stress as defined by BS 5268: Part 2: 2002 is the stress that can safely be sustained by a timber member under a particular condition. It is derived on a statistical basis from the results of loading tests on samples of timber species in controlled conditions. Structures are analysed for different loading cases to identify the maximum stresses in any structural member in order that the dimensions of the member can be selected to limit stresses to within the permissible value.

Permissible stresses are derived from timber grade stresses using modification factors (K factors) which make allowance for stress changes arising from factors such as load-duration, moisture content, load-sharing, section size etc. The grade stresses are obtained from details of the stress grading or strength class of the timber using tables 7-12a in BS 5268: Part 2.

BS 5268: Part 2: 2002 discusses design considerations (Section 1.6) some of which are applicable to general timber design and include:

- All structural members or assemblies should be capable of sustaining, with due stability and stiffness and without exceeding the relevant limits of stress, the whole dead, imposed and wind loading.

- To ensure that a design is robust and stable

    a) the geometry of the structure should be considered.

    b) required interaction and connections between timber load-bearing elements and between such elements and other parts of the structure should be assured.

    c) suitable bracing or diaphragm effect should be provided in planes parallel to the direction of the lateral forces existing on the whole of the structure.

- The design, including the design for construction durability and use in service, should be considered as a whole.

- The quality of the timber and other materials, and of the workmanship as verified by inspections, should be adequate to ensure safety, serviceability, and durability.

The standard also considers factors which condition the BS 5268: Part 2: 2002 design method and which influence the strength of the timbers in use:

- Loading – stated to be in accordance with BS 6399: Part 1, BS 6399: Part 2 and BS 6399: Part 3 and CP3: Chapter V-2 or other relevant standards.

- Accidental damage – the design should provide a reasonable probability that the structure will not collapse catastrophically because of misuse or accident. No structure can be expected to be resistant to the excessive loads or forces that could arise from an extreme cause, but it should not be damaged to an extent that is disproportionate to the original cause. It may be necessary to consider the effect of particular hazards and to ensure that, in the event of an accident, there is an acceptable probability of the structure continuing to perform its main function. In the marine and river environments, vessel impact would be one of the most likely hazards to timber structures such as river walls, walkways, and jetties.

- Service classes – because of the effect of moisture content on the mechanical properties of the timber, the permissible property values should correspond to one of the three service classes as identified in the Standard. It would be expected that Service class 3, representing external use and timber fully exposed and with an average moisture content of 20% or more would be the appropriate class for timber in river or coastal structures in most instances. A modification factor is used to derive stresses and moduli for service class 3 from those tabulated for Service classes 1 and 2.

- Duration of loading – grade stresses and joint strengths given in the Standard are applicable to long term loading. Timber can sustain greater loads for short periods and the Standard makes provision for increasing the grade stresses and joint loads for short duration loads by use of the modification factors.

- Section size – in the Standard the bending, tension and compression stresses, and the moduli of elasticity are normally applicable to members 300 mm deep (or wide for tension). Because the properties of timber are dependent upon section size and size related grade effects, the grade stresses should be modified for section sizes other than the 'basic' size as identified in the Standard, by use of the modification factors.

- Load sharing systems – the grade stresses given in the Standard are applicable to individual pieces of structural timber. Where a number of pieces at a maximum spacing of 610 mm centre to centre act together then some modification of these stresses is permitted by use of an appropriate load-sharing modification factor.

- Effective cross section – for the purpose of calculating the strength of a member at any section, the effective cross section should be taken as the target size (size used to indicate the size desired at 20% moisture content, and used for design calculations) less due allowance for the reduction in area caused by sinkings, notches, bolt, dowel or screw holes, mortices, etc. either at that section or within such a distance as would affect the strength at the section.

Limit state design

Eurocode 5: *Design of timber structures* Part 1.1: *General rules and rules for buildings* whilst specifying the rules necessary for design to the limit state method does not provide material properties as is found in BS 5268: Part 2. This information is available in other European Standards such as BS EN 338: 1995 which includes the material properties for fifteen strength classes of timber.

The limit state philosophy is based upon states beyond which the structure no longer satisfies design performance requirements. The two limit states examined are:

- ultimate limit state associated with collapse or with other form of structural failure which may endanger people
- serviceability limit state corresponding to a state beyond which specified service criteria are no longer met.

The ultimate limit states which may require consideration include loss of equilibrium of the structure or any part of it, and failure by excessive deformation, rupture or loss of stability of the structure or any part of it.

The serviceability limit states may include deformations or deflections which affect the appearance or effective use of the structure, and vibration which causes discomfort, damage, or which limits its functional effectiveness.

Characteristic values of both loads and the material properties are used for checking ultimate limit states. These are generally fifth percentile values obtained by statistical analysis of laboratory test results. Thus the characteristic strength is the strength value below which only 5% of cases lie, and the characteristic load is the load value above which 5% of cases lie. Characteristic values are modified by partial safety factors which represent the reliability of the characteristic values in particular circumstances.

The Code uses the term 'actions' to define direct loads (forces) applied to the structure and indirect forces due to imposed deformations, e.g. from temperature effects or settlement. Actions are classified and subdivided by their variation in time, e.g. permanent, variable, accidental, and by their spatial variation, e.g. fixed or free. Characteristic values for actions are modified by partial coefficient to take account of safety factors, load combinations, etc.

With regard to material properties three service classes and five duration classes are identified. Service class 3 is for a timber moisture content above 20% and for environmental conditions defined as 'timber fully exposed to the weather'. This service class is most appropriate for river and marine situations. The five load durations for actions are, instantaneous, short term (<1 week), medium term (>1 week <6 months), long term (>6 months <10 years), and permanent (>10 years). Modification factors for service class and load duration class are tabulated in the Code.

Chapter 4 of the Code deals with serviceability limit states and sets out the requirement for limiting deflection and vibration, and provides the equations for calculation. Chapter 5 deals with the ultimate limit states and sets out the design procedure for members of solid timber. Chapter 6 discusses joints and joints made with dowel type fasteners such as nails, staples, screws, bolts, etc.

The use of Eurocode 5: *Design of timber structures* is seen to provide a number of advantages including:

- the Standard represents the best informed consensus on timber design in Europe
- its widespread use will facilitate international trade and consultancy
- the use of a design method common to all materials will simplify the work of engineers
- Eurocode 5 facilitates a wider selection of materials and components than BS 5268: Part 2.

Set against the above are:

- design time will generally be greater until designers are familiar with the format
- it will be necessary to consult additional reference documents for design work.

Key considerations

The following key factors should be taken into account when designing individual members.

1. Operational Bending Moments – special attention should be given to any dynamic loadings.

2. Operational Shear Forces – again, dynamic loadings should be fully recognised.

3. Constructional Stresses – larger section sizes may be necessary to cope with constructional stresses such as driving piles into hard ground.

4. Abrasion – larger section sizes may be necessary to cope with anticipated abrasion during the life of the structure. This is critical particularly for members where replacement or maintenance would be difficult.

5. Connection Details – allowances may be necessary to compensate for loss of timber at bolt holes, bolt recesses, notches, etc. Also, allowances may be necessary to suit other connecting members.

6. Available Sizes – it is good practice to work with a limited number of section sizes for a given structure, and to industry standard section sizes. Also, the design should work within the scope of available lengths. There may be situations where it is necessary to use a larger section size in order to have available a longer length.

6.6.3. Connections

Although some preparatory work can often be undertaken, when designing connections it should be borne in mind that the majority of work often has to be carried out on site in less than ideal conditions. From a design perspective, connections can be divided into four categories: overlap, butt, scarf and notch.

Overlap

This is the simplest type of connection and involves two or more members passing across each other (Figure 6.10). In terms of straightforward design and buildability this tends to be the preferred type of connection.

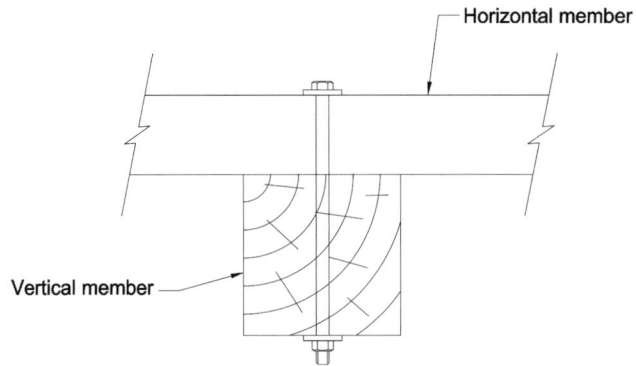

Figure 6.10.   Overlap connection

Butt

This involves two or more members converging to a single point. The members are usually within the same plane. This arrangement demands a higher degree of fit than the overlap connection (Figure 6.11).

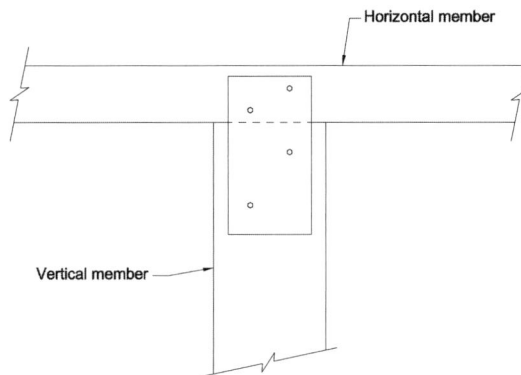

Figure 6.11.   Butt connection

Scarf

This usually involves a splice connection for the in-line extension of a member (Figure 6.12). These are more difficult to form than overlap or butt connections, so their use should be limited.

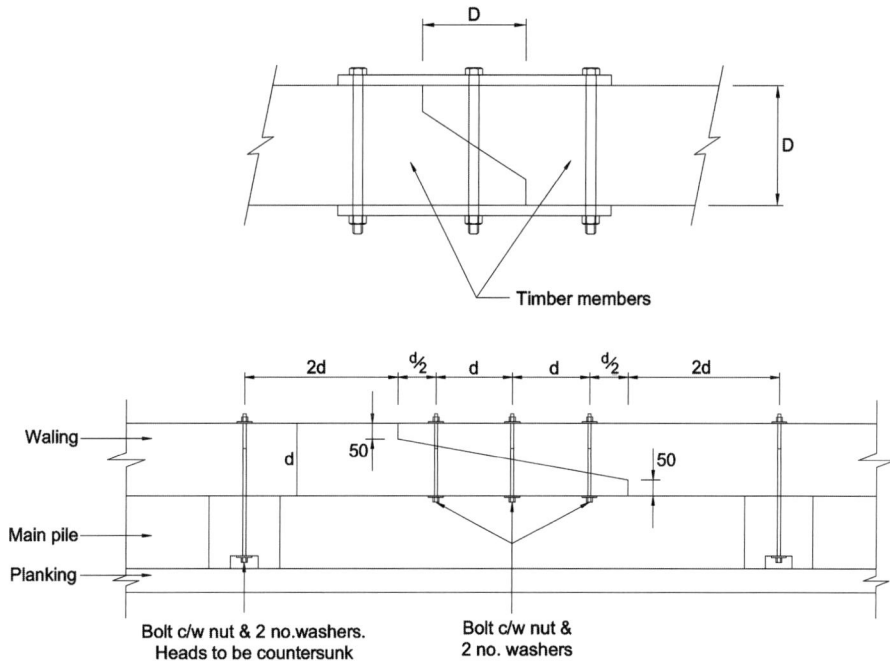

Figure 6.12.   (a) Scarf connection with metal plates (b) without metal plates

Notch

This involves cutting a notch in one timber member in order to receive a second member (Figure 6.13). Care is needed to avoid significant reduction in the size of the parent timber.

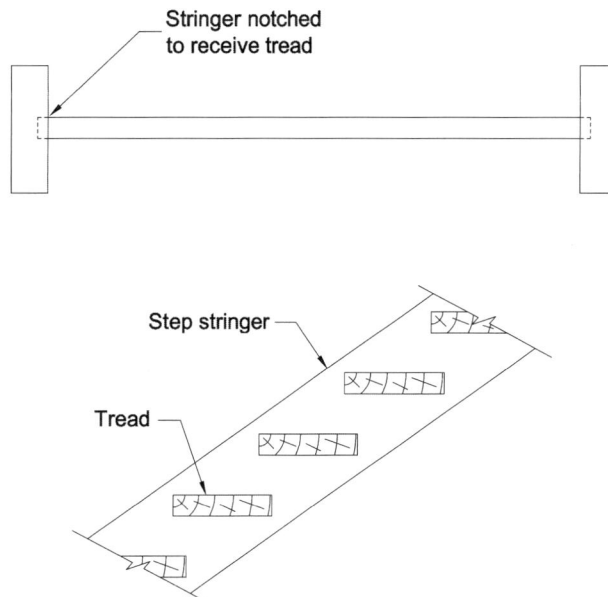

Figure 6.13.   Notched connection

Selecting connection type

When configuring the arrangement of connections within a structure, first preference should be given to designing with connections that work in compression. Namely, where the main load path is from timber to timber and the fixings (e.g. bolts) play only a secondary role.

The second preference should be for shear connections where the main load path between timbers is transmitted via the fixings working in shear. Double shear connections are more efficient than single shear connections (Figure 6.14).

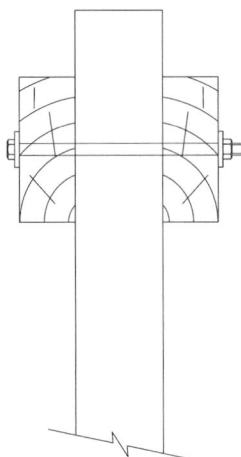

Figure 6.14.   Double shear connection

The third preference should be for tension connections where the main load path is via fixings working in tension.

Other points to bear in mind are:

- timber interfaces at connections should be in full contact with no significant gaps
- as a general rule no packing pieces, shims or similar should be used in connections
- any shaping or trimming of members at connections should avoid producing thin sections and 'feather edges'
- any notching of timber at connections should be kept to a minimum
- where possible connections should be staggered within a structure to avoid lines or planes of weakness.

6.6.4. Fixings

The traditional mechanical fasteners for structural timber are divided into two groups depending on how they transfer the forces between the connected members. The first group includes dowel-type fasteners, bolts, coach screws, dowels and nails. Here, the load transfer involves both the bending behaviour of the dowel and the bearing stresses in the timber along the shank of the dowel. The second type includes fasteners such as split rings, shear-plates, and punched metal plate fasteners where the load transmission is primarily achieved by a large bearing area at the interface of the members.

Fixings should be designed to be effective for the design life of the structure through the use of non-corrodible materials appropriate protection systems for the fixings, or by making due allowance for the corrosion that will occur over the life of the fixing by over sizing the section used.

Bolts

Bolts are threaded dowel-type fasteners, with hexagonal or square heads and nuts. When installing a bolt, the pre-drilled hole must include a clearance that allows for easy insertion of the bolt and also timber movement so that any tendency for the members to split on assembly or after drying out is reduced. The clearance must be carefully selected to provide best performance for given materials and operating conditions. BS 5268-2 allows a diameter up to 2 mm larger than the bolt diameter, but Eurocode 5 allows a maximum difference of 1 mm. Bolts should be tightened so that the members fit closely and should be re-tightened when the timber members reach equilibrium moisture content.

For most situations bolts are the preferred fixing as they provide a positive and robust connection and are relatively straightforward to fix. Galvanised, coarse threaded, mild steel bolts are usually the most cost effective solution but other types such as stainless steel are more durable, facilitating easier removal (i.e. in dismantling or repair of the structure) and reuse.

Whenever a bolt is used, either on its own or through a connector, it should be installed with a washer under any heads or nuts which are in contact with the timber. Timberwork washers should have a diameter of a least 3 times the diameter of the bolt and a thickness of 0.3 bolt diameters to provide an adequate bearing area. Bolt heads and nuts in contact with steel surfaces follow normal washer rules for metalwork. Where, in current or future works, timbers will pass across the end of a bolt, the head of the bolt or the nut (as applicable) should be countersunk into the timber to avoid conflict and ease construction.

As a general rule, large diameter bolts are to be preferred: 16 mm diameter should be seen as a minimum for, say, lightweight connections, and for heavy duty connections 30 mm diameter may be appropriate. For most situations 20 mm and 25 mm diameter bolts tend to be suitable. Larger diameter bolts have an advantage in terms of exerting lower secondary stresses on the timber and coping better with corrosion.

Within a connection, bolts should not be designed to bridge an air gap. The length of bolt between the head and nut should be fully enclosed. Close attention should be also be paid to the edge distances between the bolt holes and the timber faces and the spacings between bolts.

Coach screws

Coach screws are also threaded dowel-type fasteners with hexagonal or square heads, but the thread is coarser with greater pitch and has a cone point enabling it to cut into and grip the timber directly. Most coach screws have an unthreaded length of the shank adjacent to the head enabling the cladding to be held tight against the parent timber when the screw is turned home. They are normally limited to fixing cladding such as planking and decking, and are suitable for multiple lightweight connections but installation requires a higher degree of workmanship than bolts. Typically, they are galvanised mild steel, 16 mm in diameter with a 25 mm square head (Figure 6.15) although other materials may also be used. To ensure maximum grip, coach screws should be wound home (normally to a specified minimum torque) and not struck home.

In situations where it is likely to be necessary to remove fixings for operational or maintenance purposes, coach screws are less suitable than bolts. Their heads are less robust than bolt heads and their grip within the parent timber is susceptible to deterioration through repeated fixings.

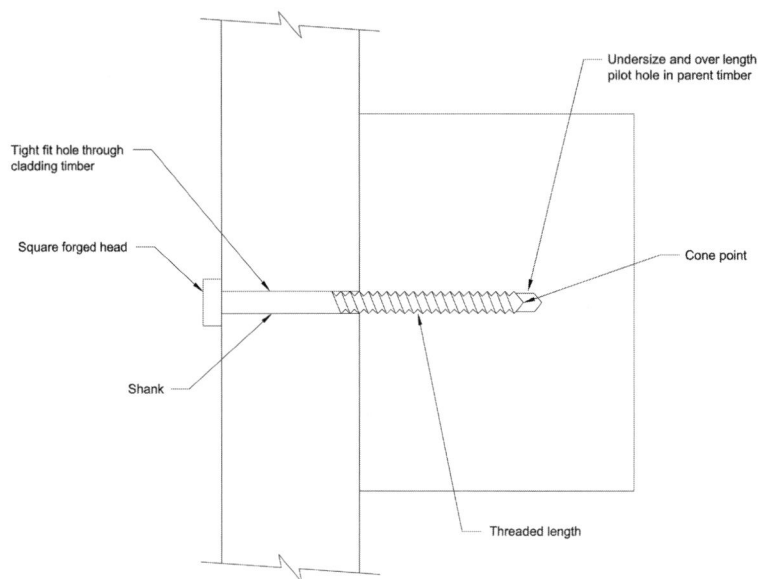

Notes:

1. Coach screws used in various lengths and diameters.

2. 16 mm dia. x 175 mm long is the largest available 'off the shelf' and is the one usually used in heavy duty groyne construction

Figure 6.15. Coach screw fixing

Dowels

Dowels are cylindrical rods, generally of smooth surface, available in diameters typically from 6 mm to 30 mm. They are inserted into pre-drilled holes with a diameter not greater than that of the dowel itself. This means that the holes must be accurately positioned, which is generally achieved by drilling through one member into the other.

Timber dowels can be used but are rare due to the high standards of workmanship required. Their main advantage over steel is their ability to expand, contract and deflect in harmony with the parent timber. For these reasons such connections often remain tighter for longer periods of time and can be particularly appropriate for structures experiencing high impact loads such as mooring jetties.

Joints made with dowels have a better appearance than bolted joints and are also stiffer. If steel members are incorporated in a dowelled joint, the holes in the steel members must include a clearance, and due allowance should be made for any extra slip that may occur as a result. The use of dowels may not be appropriate for large timber structures as timber drying may exert high stresses on the structure. However, if the structure remains in the green state and the timber does not dry, dowels may be considered.

In large dowelled connections it may be necessary to replace some of the dowels with threaded bolts to stop the joint from opening laterally.

Nails, spikes and timber dogs

Heavy duty nails etc. may be used in limited situations such as for decking and causeway timbers. However, they are less suitable for resisting tensile loads and they cannot be easily removed if the need arises, thus coach screws are usually preferred in these situations. Particularly with the larger sizes and thin timbers, pilot drilling may be required to prevent splitting of the wood.

Shear connectors

Although popular in other applications toothed plates, shear plates, split ring connectors, etc. are rarely used in coastal and river engineering due to corrosion and the difficulties associated with forming connections in situ, especially with hardwoods.

Glue

Glued connections are generally impractical, especially as stand alone connections. However, bolted and dowelled joints may be made stronger by incorporating a resin into the joint.

6.6.5. Ancillaries

There are a number of ancillary devices that can be used to facilitate or strengthen a connection.

Steel washers

These are normally used with bolts in order to reduce the local bearing pressures exerted by the bolt heads and nuts on the timber. To provide effective load distribution and to cope with corrosion the washers need to be of the order of 10 mm thick. Typically, washers are in galvanised mild steel and diameters of 50 mm, 75 mm and 100 mm are used depending on the bolt size and tensile loading.

It is not normal practice to use washers with coach screws except where they are used with soft woods (such as Douglas fir) where it stops the head of the coach screw becoming overly embedded in the timber. If there is an issue with significant pull-out loads then it is usually advisable to use bolted connections instead of coach screws if at all possible.

Steel flat plates

Again, these are normally used with bolts. The main purpose of plates is to increase the capacity of the connections in their ability to transfer load from one member to another. Usually they are used in butt joints and scarf joints.

Plates can be used singly or in pairs. Where pairs are used the first plate has close fit bolt holes and the second has oversize holes to allow for drilling tolerances through the timber. Pairs of plates are more efficient than single plates (due to the advantages of double shear) but their installation demands higher standards of workmanship. Also, the greater the number of bolt holes the more difficult the installation.

Plates are typically in galvanised mild steel and of the order of 12 mm thick (Figure 6.16).

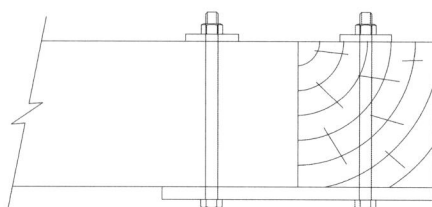

Figure 6.16. Single plate connection   *Figure 6.17. Steel angle connection*

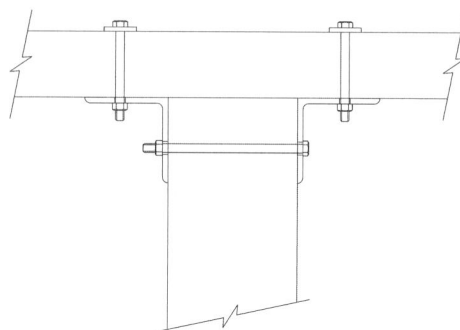

Timber plates

In some situations timber plates can be used in place of steel ones. To achieve similar strength characteristics they need to be of much greater thickness and possibly longer; however they do not normally require prefabrication, and it is one way of using up offcuts from site fitted timbers.

Steel angles

Their application and detailing are similar to flat plates except that they are mainly, but not exclusively, used in overlap joints (Figure 6.17).

Steel straps

These are broadly similar to plates and angles except for being more versatile in as much as they can be bent to various shapes, including curves. Also they lend themselves to being wrapped around members thereby reducing the number of positive bolt or coach screw fixings required.

Spliced piles

These should be avoided where possible but may be needed where, for instance, plant bays are required. The design issue is to ensure the transmission of full bending moments across the spliced joint. Figure 6.18 gives a possible detail to solve this problem.

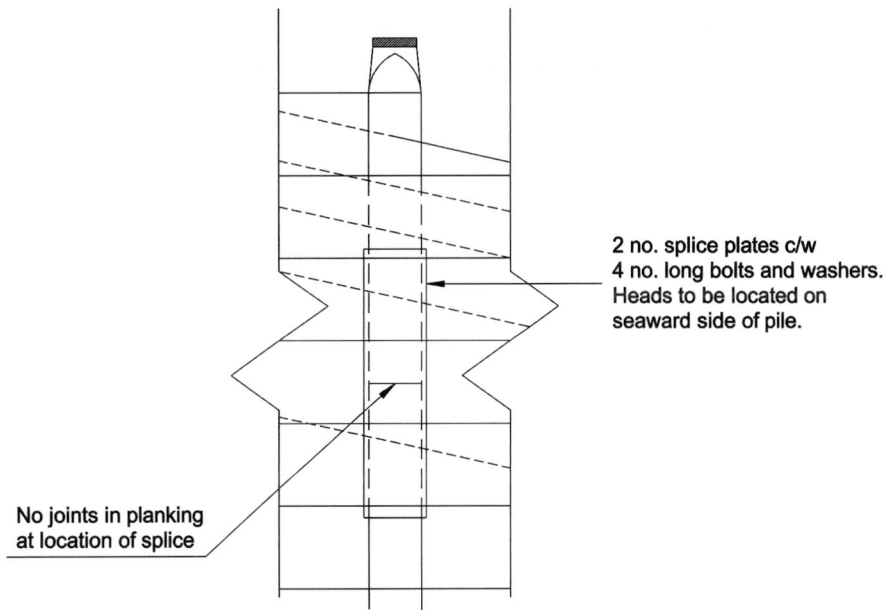

Figure 6.18.   Spliced pile detail for a groyne

The technique may be modified if piles need extending during driving, when either four plates or a complete collar is used. In this instance, the receiving pile head must be cut off horizontally and the extension piece squared off to match it; any failure to make the joint in the horizontal plane or to achieve exact alignment across the joint will lead to its failure.

6.6.6. Connection design

The design of joints using dowel-type fasteners should be undertaken in accordance with BS 5268: Part 2 or Eurocode 5 (EC5). Although it remains a permissible stress design code, BS 5268: Part 2 now takes the EC5 approach to the design of screwed, bolted and dowelled joints. The terminology and notation used here is largely based on Eurocode 5, but the principles in all the above publications are very similar.

Spacing rules

Dowel-type fasteners must be spaced at suitable distances from each other and from the ends and edges of timber or wood-based materials to avoid splitting. An end distance is said to be loaded when the load on the fastener has a component towards the end of the timber, otherwise it is referred to as an unloaded end distance. Loaded end distances need to be greater than unloaded ones. Edge distances may be loaded or unloaded in a similar way.

The values for spacings and distances vary from one fastener type to another, as well as between the various material types. Factors taken into account in producing the code recommendations include the cleavage and shear strength of the timber, the timber density, the fastener diameter and, in some cases, the angle of load to the grain. The spacings and distances recommended in the codes are based upon testing and years of experience. Deviations should not be made unless strongly supported by tests and by an independent authority.

In most joints made with coach screws, bolts or dowels, the fastener is loaded in lateral shear. To calculate the load-carrying capacity of such a joint, values for three material properties are needed. These are the timber density, the embedment strength of the timber or wood-based material and the bending strength of the fastener.

EC5 provides formulae, based on test results, for calculating the characteristic embedment strength of commonly used fastener/timber combinations from the density of the timber and the diameter of the fastener. Similar formulae are included in an Appendix to BS 5268: Part 2: 2002. In the case of EC5, the characteristic embedment strength is converted to a design value by applying the appropriate load duration factor and the partial safety factors for timber and connectors. The tables of basic loads for fasteners in BS 5268: Part 2: 2002 have been derived using the same principles as EC5, but with load-carrying capacities adjusted to match the permissible stress design basis of this code. Design rules are given in Section 6 of BS 5268: Part 2 for use in determining permissible loads.

The characteristic bending strength of the fastener is calculated from formulae given in EC5 and converted to a design value by applying a partial safety factor for the fastener material. In the BS 5268: Part 2: 2002 tables, fastener strengths are similarly taken into account. For non-standard and proprietary types of fastener or wood-based material, test methods for determining the embedding strength of the fasteners and the yield strength of dowels are taken from EN standards.

For laterally loaded joints, there are numerous possible modes of failure and the strength of these joints can be calculated using Annex G of BS 5268: Part 2 or EC 5, provided that the strength of the steel from which the coach screw is made is known.

Slip

Slip between the component parts occurs to some degree in all mechanical joints as the joint takes up load. It can be a significant factor in the deformation and, in some cases, the strength of some components. BS 5268: Part 2 and EC5 both include calculation methods for slip in joints. The magnitude of slip is time-dependent and is also affected by fluctuating moisture content in the timber.

6.7.   FINISHES AND FITTINGS

The final stage in the design process is the design of finishes and fittings, however it should be noted that there are instances in which the fixtures and fittings will influence other aspects of the design process, requiring further iteration in design calculations.

6.7.1. Finishes

Because of the often aggressive environment in which coastal and river timber structures operate, the range of appropriate finishes tends to be limited. As a general rule durable timbers are left in their original (often as-sawn) condition. Where the application requires a smaller tolerance (such as decking, where even small differences in levels between planks could form a trip hazard), the timber may be machined to rounded edges or particularly smooth finish. It should be noted that many of the tropical hardwoods used in coastal and river engineering are difficult to work, and the finish required could influence the selection of timber.

Machining timber, including cutting grooves in deck timbers to improve their surface grip or rounding edges of hand rails, will inevitably reduce the section size and this should be considered when designing members. Machining is usually undertaken off site prior to the delivery of timber. Proprietary manufactured timber boards with factory-applied non-slip coatings are available for use where timber surfaces would form a slip hazard.

6.7.2. Fittings

Fittings tend to facilitate either construction or operational activities, but some have a dual purpose.

Construction facilities

These are fittings which primarily aid construction, such as pile rings and pile shoes (Figure 7.11). Items of this nature need to be properly sized for their function and carefully fitted. Poor workmanship can result in such fittings becoming a hindrance rather than a help.

Operational facilities

These are fittings which primarily aid the operational use of the structure such as mooring bollards, fenders and marker beacons. They can also include operational equipment such as davits. Fittings of this nature invariably introduce their own loadings to the structure and these need to be fully taken into account in the design. Such loadings can influence the overall stability of the structure and the foundation arrangements.

7. Design and construction issues

Whereas the previous chapter dealt with the main mechanics of the design process, the following text describes the fine tuning of designing and constructing a competent timber structure. Whilst they may appear minor, good detailing and construction practices can significantly enhance the performance and service life of timber structures, as well as making them easier, safer and more economic to build.

7.1. HEALTH AND SAFETY

All parties have a responsibility to ensure that construction activities are safe and undertaken in accordance with established best practice. Under the Construction (Design and Management) Regulations (CDM, 1994), particular responsibilities are placed on each party. Those allocated to the designers are described in Box 7.1.

Box 7.1.   Designer's responsibilities under the CDM regulations

The role of the designer is to design in such a way as to eliminate or reduce risks to the health and safety of those who are going to construct, maintain, repair, demolish, or clean any type of structure. The remaining risks should be as few and as small as possible. The overall design process need not be dominated by the need to avoid all risks during construction and maintenance but to reduce unnecessary levels of risk.

The designer's key duties under the CDM regulations as far as is reasonably practical are:

- to alert clients to their duties (particularly with regard to appointing a planning supervisor)
- to consider, during the development of designs, the hazards and risks which may arise to those constructing and maintaining the structure
- to design to avoid risks to health and safety so far as is reasonably practical
- to reduce risks at source if avoidance is not possible
- to consider measures which will protect all workers, if neither avoidance nor reduction to a safe level is possible
- to ensure that the design includes adequate information on health and safety

---

Box 7.1.   Designer's responsibilities under the CDM regulations (continued)

- to pass this information on to the planning supervisor, principal contractor, or others who might need it, so that it can be included in the health and safety plan and health and safety file
- to co-operate with the planning supervisor and other designers.

7.1.1. Design risk assessment

Design risk assessment is the examination of ways in which hazards can be avoided or mitigated, or, if neither is possible, designed so that the level of risk is acceptable by applying the principles of prevention and protection.

The process whereby a designer can logically identify, assess and manage risks is as follows:

- identify hazards and assess the risks arising from them
- change the design to eliminate hazards, or to reduce risks where hazards cannot be eliminated
- pass on information to others.

Designers will have to assess a likely scenario by using their experience, imagination and professional judgement. This will be informed by an appreciation of the hazards and risks to be avoided, mitigated, or accepted and controlled.

7.1.2. Construction safety

A pre-tender Health and Safety Plan should be made available at time of tender which identifies the significant Health and Safety hazards and risks (Figure 7.1), and the standards to be applied to control the significant Health and Safety risks.

Figure 7.1. Access and interface with the public during reconstruction of timber groynes at Eastbourne, East Sussex (courtesy Mackley Construction)

In the context of work involving construction with baulk timbers potential hazards which may require consideration involve:

- storage of timbers on site
- lifting and handling of timbers
- the use of machinery for cutting and drilling timbers
- application of timber preservatives.

7.1.3. Potential hazards for marine or river construction work

Some potential hazards that may need consideration during the design process are included in the following table: other information may also be available from Simm and Cruickshank (1988) and Morris and Simm (2000).

Table 7.1. Potential hazards for coastal and river construction

Potential hazard	Mitigation measures	Outcome
1. Lack of robustness and/or stability in the design particularly during the construction process	Designer should check the design to ensure partially built structure is stable. Designer should state in Health and Safety Plan whether any special precautions or propping are necessary at each and every stage of the construction process to ensure overall stability of all parts of the structure.	Hazard reduced or removed
2. Structure is a hazard to navigation	Designer should consider including navigation lights and/or marker posts in the design. Fenders should be included at positions where boat impacts are likely.	Hazard reduced
3. Construction involves working over water	The design should maximise the amount of work that can be carried out from land. The design should be simplified to allow quick and easy erection.	Hazard reduced
4. Construction involves intertidal work (Figure 7.2)	The design should minimise the time needed for work between tides. Consider prefabrication of components where possible and avoid use of in situ concrete.	Hazard reduced
5. Maintenance work involving the replacing of damaged/deteriorated structural members	The design should identify those members which are likely to need replacing and ensure that they can be removed and replaced easily.	Hazard reduced
6. Cleaning, maintenance, or working personnel slipping and falling into water	The design should include the provision of life belts, ladders and grab chains. Working decks should be finished with non-slip surfaces. Handrailing should be provided where work near a drop is necessary.	Hazard reduced or mitigated

Figure 7.2. Groyne construction work at Eastbourne, working up the beach with the rising tide (courtesy Mackley Construction)

7.1.4. Working with timber

Information concerning the toxicity of timber is published by the Health and Safety Executive (2003). Timber itself is unlikely to cause a toxic effect, although the dust, sap, resin, latex, or other organisms on the timber such as lichens and moulds can cause toxic reactions. The carcinogenic effects of dust from certain hardwoods has been well documented. Toxic activity is species specific, therefore knowing the species is important in establishing what the potential effects may be.

The reaction to different species of timber is heavily dependent upon the sensitivity of the person exposed to the timber. Exposure to dust, for example, can result in one or more of the following symptoms and the severity of the symptoms can vary:

- dermatitis
- rhinitis
- nosebleeds
- nasal cancer (this is rare and tends to be triggered by long term exposure to dust)
- asthma
- impairment of lung function
- extrinsic allergic alveolitis (rare condition resulting in progressive lung damage)
- conjunctivitis
- sore eyes.

Once exposed to certain timbers, the individual can become sensitised, therefore, further exposure can result in adverse reactions.

There is a commonly held misperception that splinters from certain species of timber can result in septicaemia. It is the bacteria and/or fungi on the splinters that are responsible predominantly for infection and subsequent poisoning and not the timber.

The potential ill effects of working with timber are easily avoided by adopting good health and safety procedures. When working in an enclosed space, there should be an effective dust extraction system and appropriate personal protective equipment such as dust mask, goggles and overalls covering exposed skin should always be used.

7.2. ENVIRONMENTAL IMPACTS AND ENVIRONMENTAL APPRAISAL

The identification, measurement, and assessment of environmental impacts is known as 'environmental appraisal'. Assessment in this context refers to the process of determining the importance of any impact. Thus an impact might be identified as the effect, over the lifetime of the scheme, of constructing new timber groynes on a site of nature conservation importance. This impact might be measured in terms of the change in the area, or the quality, of habitats. Assessment involves determining the importance of these impacts.

7.2.1. Environmental impact in the appraisal process

Ideally environmental appraisal should start with appraisal at the strategic level to reduce the risk of unacceptable options being taken forward. The strategic level appraisal will allow the combined effects of schemes to be considered.

Appraisal should take place through the whole scheme development, starting at inception and continuing through option development and choice, scheme design and operation, audit, and, finally, post project appraisal. It is important to ensure that environmental appraisal interacts with economic appraisal so as to ensure that the final design is both environmentally acceptable and economically viable.

To achieve environmentally acceptable results when designing a scheme, in conjunction with their clients, designers should:

- seek to avoid environmental damage
- minimise environmental damage where some impacts are unavoidable
- devise suitable mitigation measures to offset residual impact where possible
- identify and, where practical, include opportunities for environmental enhancement.

It is necessary to consider environmental impacts throughout the life of a scheme. For example, the build-up of sediments in new areas is likely to have long-term effects on landscape and nature conservation. The timing of construction works should also be considered since this can often change environmental impacts. Where an initial consideration shows an option likely to be particularly environmentally damaging then it should not normally be taken further.

Minimising the risk to features of interest in designated sites is critical. In considering options it is important to examine the reasons for designation. Where a site is designated for species associated with wetland habitats, then lowering water levels is unlikely to be acceptable. The advice of all interested organisations should be sought when considering appropriate options for designated sites.

Opportunities for reducing or eliminating environmental impacts, and/or providing environmental enhancements should be investigated at the detailed design stage. An example would be the selection of timber for the construction of a flood defence wall to blend better with the landscape and to provide habitat opportunities for certain types of flora and fauna.

The medium to long term benefits of environmental protection and enhancement measures must be considered against costs at this stage. For example, it will not always be cost effective to save every tree along the line of a flood defence wall. There may be cases where the environmental benefits of a sensitive mitigation scheme outweigh the benefits achieved at greater cost through avoidance measures.

7.2.2. Legal requirements for environmental appraisal

The majority of timber structures in the coastal and river environments have a flood or coastal defence function. In these instances an environmental appraisal must be carried out prior to application for grant aid. In certain circumstances the environmental appraisal must be in the form of an Environmental Impact Assessment which will include the preparation of an Environmental Statement. The Environmental Impact Assessment (Land Drainage Improvement Works) Regulations 1999 apply to flood defence capital works if they are improvements to existing defences but new flood defence and coast protection works are subject to the Town and Country Planning (Environmental Impact Assessment) (England and Wales) Regulations 1999. These Regulations implement Council Directive 85/337/EEC and its amending directive 97/11/EC, on the assessment of certain public and private projects. An Environmental Statement is required where the impact of such projects is likely to be significant.

7.2.3. Factors for consideration in environmental appraisal

In order to fully identify the likely significance of different scheme options when undertaking an environmental appraisal, sufficient environmental information must be obtained by a combination of reviewing existing information and collecting new data.

The Flood and Coastal Defence Project Appraisal Guidance – Environmental Appraisal (FCDPAG5) produced by the Department for Environment, Food and Rural Affairs (previously MAFF) identifies a series of environmental factors that are particularly relevant to the assessment of flood and coastal projects.

These factors are:

- flora
- fauna
- population
- cultural heritage

- property and the built environment
- landscape
- geological/geomorphological features.

Scheme or option impacts on these factors would need to be thoroughly considered. Other effects that should also be considered are:

- the impact of construction traffic
- impacts on access (e.g. public footpaths)
- impacts due to construction noise and air emissions
- longer term recreation and amenity considerations
- social and economic considerations
- health and safety (especially during construction)
- impact on soils and mineral deposits
- water quality implications (both surface and groundwater).

7.2.4. *Environmental design considerations*

In coastal situations timber is used among other things for breastworks, groynes, piers, steps, ramps, boardwalks and as stepped reinforcement or fencing protection to dunes. In river situations timber is used for bank retention/reinforcement, ramps, steps, boardwalks and fishing platforms.

Weathered timber is an important component of beaches, harbours and rivers where it often reinforces the traditional landscape character. The painter John Constable wrote of his native River Stour: *"the sound of water escaping from Mill dams,....Willows, old rotten banks, slimy posts and brickwork. I love such things...They made me a painter and I am grateful"*. Similarly, timber groynes, sculpted by the sea and part buried in the sand are a classic element of many beach scenes, and it is for this reason that the seaside towns of Bournemouth and Folkestone have deliberately continued to promote their use over, say, rock armouring or concrete.

Along the coast a number of good practice principles and/or problems frequently occur and should be considered whenever timber is used.

- As a gentler material than steel and concrete, timber is often more attractive for use in areas of high amenity and its use should be considered therefore where possible. However, it should be part of a carefully designed scheme which builds on the existing sense of place and takes all available opportunities to mitigate undesirable impacts.

- Many such areas along the coast or close to rivers are nature reserves or managed for conservation or recreational use. Liaison with those responsible for these areas should be a first step and involvement with local wardens or staff may be particularly helpful since they will often be involved in any maintenance and/or repair activity.

- New structures should not be built or altered in protected areas such as sensitive marine habitats or protected landscapes without proper consultation with relevant statutory bodies. Structures may alter the natural processes such as sediment transport or deposition which, in turn, may affect the local fauna and flora. Archaeology will be an important issue especially in the intertidal zone where permanent saturation preserves ancient artefacts such as shipwrecks or the prehistoric Seahenge discovered on the Norfolk coast in 1998.

- Groynes need to be built to last as long as possible but visually sympathetic detailing such as strong use of verticals and a slightly uneven top line should also be considered. It may be useful to consult a landscape architect or architect at the design stage. When a new series of groynes is erected, the old groynes, which may have weathered into attractive features as well as providing shelter on a windy beach, should not be automatically removed. Ecologists should be consulted in relation to the possible impact of treated wood on marine life.

- Sand dunes are often valuable habitats, while also acting as an important part of the sea defences. They do however often get eroded by holiday-makers. Directing pedestrians along timber boardwalks and steps is therefore often desirable and has been achieved successfully at a number of locations. Sweet Chestnut fencing is also sometimes used to prevent people eroding dunes and also disturbing nesting seabird colonies. There are excellent reasons for using locally grown chestnut in the south of England since regular coppicing preserves the woodland habitat. There are very few suppliers. Information on the usefulness of this fencing in protecting the coastal landscape should be made more available to the public since it is sometimes taken to supply beach barbecues.

- Boardwalks should be designed with a careful view to their function, safety, maintenance and appearance. Joists make decking cheaper by allowing greater spans but they may need bracing. Beams need to be spaced 1.8 to 2.4 m apart if there are no joists or 2.4 to 4.9 m apart if joists are used. Posts can be projected up for handrail supports and should be square sectioned to prevent twisting. Gaps between decking boards must be no greater than 15 mm otherwise they will form a trip hazard. Wood can be slippery and safety is of paramount importance especially near water; non-slip treatments, or proprietary coated wood panels should be considered.

- Timber steps constructed over earth sea defences cause less permanent damage to the defences than steps made out of materials such as concrete. Steps should not be too narrow and of leisurely ascent.

- Surfaces on timber steps and ramps present the same safety considerations as timber decking on boardwalks. Whilst exposure to sun, wind and salt makes timber less prone to slipperiness in coastal situations than in riverine situations, steps which are regularly inundated by high tides are very slippery.

7.2.5. *Actioning the outcome of environmental appraisal*

An environmental appraisal having been undertaken, it is important to ensure that its recommendations are incorporated into the final design and construction process. A method of doing this is to prepare an Environmental Action Plan which will set out environmental actions to be undertaken before, during and after construction together with the method of managing and monitoring its implementation.

7.3. DURABILITY

The durability of timber structures is key to minimising environmental impacts and whole life costs. It maximises the lifespan of the structure and/or extends the maintenance intervals by, among other things, details such as:

- provision of sacrificial wearing plates or oversize members at locations of high abrasion
- avoidance of local sediment traps, passageways and other configurations which might encourage high wear and tear through the channelling of water and/or water-borne sediment
- avoidance of traps or surfaces which will result in the collection of water resulting in rot
- avoidance of connections which are reliant on single fixings and/or situations where a simple failure would allow the structure to 'unravel'.

7.3.1. *Durability by design*

No construction is capable of lasting forever. Nevertheless, structures built of timber are capable of achieving long and effective service lives, provided that the correct principles are considered. With time, moisture and dirt can become trapped at member junctions unless good details are provided to minimise the risk at such locations. Trapping points lead to premature biological attack which then spreads to other parts of the structure. Careful detailing provides conditions for moisture drainage and air circulation at such points.

Section 9.1 describes the decay process that may occur in both river and marine environments. Essentially, timber will not rot if it is kept below a moisture content of 20%: this is commonly referred to as the decay threshold. Generally, timber that is serving in a Hazard class 3 situation will periodically attain a moisture content above the decay threshold. Timber serving in more onerous hazard classes may be permanently at risk, such as those in ground contact or immersed in water. However, it is important to appreciate that timber immersed in water is most at risk at the water/air interface.

By appreciating the relationship between moisture content and fungal decay, it is possible to enhance and prolong the service life of timber structures by following a few basic tenets. The following four principles illustrate the key points for improving the durability of timber structures serving in Hazard class 3 (HC3). In other words, achieving durability by design.

1. Provide effective drainage from timber and ensure the surfaces are well ventilated.

2. Protect exposed end grain and avoid water traps and capillary paths.

3. Protect the top of horizontal members.

4. Avoid direct contact with other absorbent materials.

Good drainage details minimise the volume of water that can accumulate on horizontal surfaces. Furthermore, by appreciating the effects of surface tension, water management design details can be improved so that the risks of fungal decay are minimised. Figure 7.3 exemplifies the principles of durability by design by providing good drainage away from horizontal surfaces: a minimum fall of 1:8 is recommended.

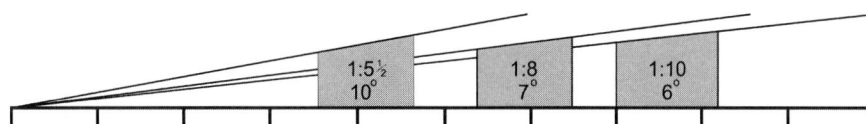

Figure 7.3. Typical 'fall' of horizontal surfaces (courtesy TRADA Technology)

Sharp corners tend to break surface tension (Figure 7.4) whereas radiused corners maintain surface tension. When combined with a 'fall' of 1:8, water is drained away from the horizontal surface. The water on the vertical face pulls the water off the horizontal face. The radiused surface should be a minimum of 3 mm.

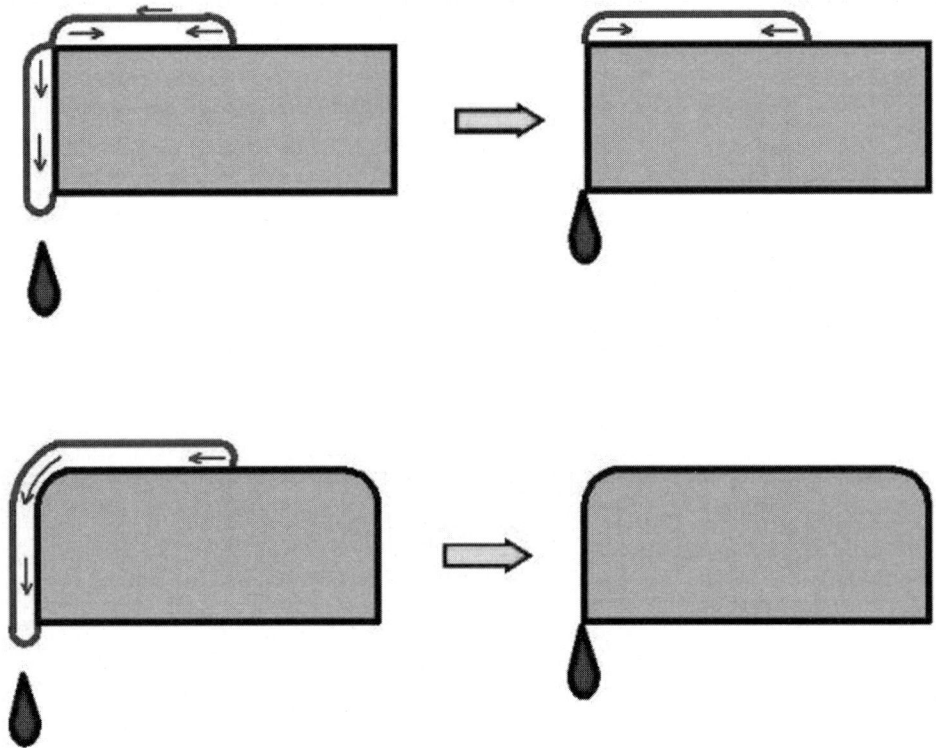

Figure 7.4. The advantages of incorporating a radiused corner to a horizontal surface (courtesy TRADA Technology)

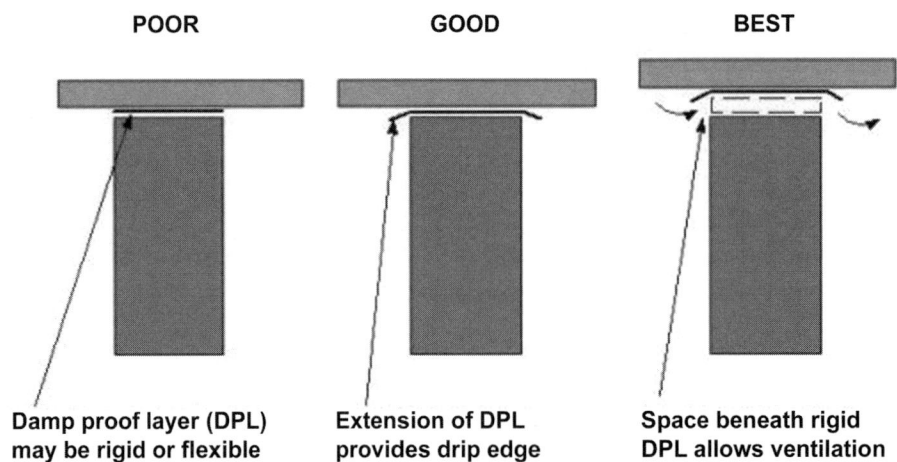

POOR **GOOD** **BEST**

Damp proof layer (DPL) may be rigid or flexible

Extension of DPL provides drip edge

Space beneath rigid DPL allows ventilation

Figure 7.5. Good ventilation over a horizontal surface (courtesy TRADA Technology)

In Figure 7.5, the horizontal top edge of the beam is offered some protection by the damp proof layer. However, this protection is optimised by extending the DPL to provide a drip edge and also by incorporating a spacer to allow ventilation of the joint. Good airflow minimises the duration where conditions can result in the moisture content exceeding the decay threshold of 20%. There may be some parts of timber constructions where it is difficult to incorporate adequate drainage and ventilation, therefore specialist advice should be sought concerning the 'armouring' of such details with a localised wood preservative treatment.

One of the most common causes for premature failure in external timber structures is the lack of a capping detail. The microscopic structure of timber may be simply described as a series of 'drinking straws'. Therefore water can penetrate down exposed end grain faces with the result that the consequent increase in moisture content can support fungal decay. By capping the end grain of exposed elements, the risks of fungal decay penetrating along the wetted end grain are significantly reduced.

Figures 7.6 and 7.7 provide good examples where the end grain of glulam components have been protected. Figure 7.7 illustrates the drip detail which discharges water away from the faces of the beam.

Figure 7.6. Protecting end grain (courtesy TRADA Technology)

Figure 7.7. Protecting end grain and providing a drip detail (courtesy TRADA Technology)

Figure 7.8. Plan view of a greenheart pile approximately 100 years old. The outer perishable sapwood has eroded, leaving the cylindrical durable heartwood clearly visable (courtesy TRADA Technology)

Many marine pilings lack a capping detail. Despite being situated in a hazardous environment, the most common mode of failure is by fungal decay penetrating down the component above the water line. Moisture penetrating down through the exposed end grain can elevate the moisture content within the pile above the decay threshold of 20% and thus trigger fungal decay. The decay will tend to occur more rapidly in perishable sapwood, if present on durable hardwood piles as illustrated in Figure 7.8, or in the less well preservative-treated core of softwood piles. The service life of a pile can be enhanced by including a capping detail (Figure 7.9). In the case of hardwood piles, a sheet of corrosion resistant metal may be used, or, in the case of preservative-treated cylindrical softwood piles, proprietary pile caps may be used. The traditional use of steel rings applied during or after the driving of piles (as described in Section 7.6) may also provide some protection through compression of the end grain.

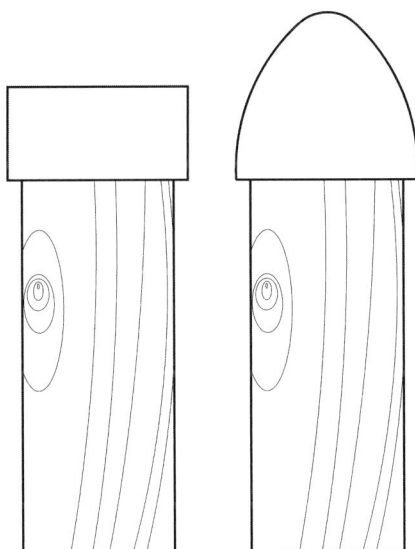

Figure 7.9. Caps for marine piles

As stated earlier, it may not always be possible to fully protect joints in timber structures and it is inevitable that moisture will remain trapped and localised areas of high moisture content may occur. For example, the joint where a supporting beam connects with a piling in a bridge may be particularly at risk. It may not be structurally practicable to introduce ventilation around the joint. Therefore, a drainage channel could be drilled into the ventral sector of the joint. Additionally, localised wood preservative such as boron-containing rods or gels could be added to the joints to provide preservative protection against fungal decay.

Boron-containing rods or gels are highly concentrated reservoirs of boron-based wood preservative. The preservative is activated and mobilised by moisture, in other words, the preservative is only activated when conditions are conducive for fungal decay. Boron-based preservatives are also recognised for their low environmental toxicity. However, immersion in water should be avoided as such preservatives can leach out into the environment, thereby also reducing their efficiency. Boron-based preservatives should only be used in Hazard class 3 and specialist advice should be sought before designing for localised preservative treatment.

Box 7.2.   Durability of groynes at Bournemouth

Bournemouth has a field of 50 impermeable timber groynes, typically 75 m long at 180 m spacing. The piles are 300 mm × 300 mm, and the planks are 300 mm × 100 mm, and predominantly of greenheart. Six groynes were planked with ekki, one with balau, and one with jarrah in the mid 1980s. Aluminium bronze bolts were used in 1971-73, and stainless steel thereafter. In recent years coach screws have been used in preference to bolts. The beach is a fine sand, with an appreciable shingle content.

The groynes typically have a serviceable life of 25 years. Abrasion of the timber is negligible if the groyne remains impermeable (with wear of less than 25 mm). The stainless steel bolts can usually be undone (and can sometimes be reused) after 25 years' service. Wear is usually confined to the exposed threads.

Stainless steel bolts cannot be done up as tightly as galvanised bolts because of a tendency for grit to jam in the threads. The use of a second lock nut is essential, but even then some work loose. Coach screws are now preferred because:

- they can be done up more tightly
- there is no exposed thread to risk injury to beach users
- one less man is needed during installation
- less excavation is needed at the rear of the groyne.

Four permeable groynes were built to three different designs. All showed major abrasion as sediment laden water swirled around piles and all have been replaced (or soon will be). Permeable groynes:

- are no cheaper than impermeable groynes
- are ineffective
- abrade at a faster rate
- need more maintenance.

Gribble infestation is the major cause of damage. It is often confined to a few timbers in any one groyne, with adjacent timbers being completely unaffected. Timber infested with gribble is more liable to abrasion, and gaps open up between planks, leading to concentrated water and sediment flows, and high abrasion. The loss of retained fill is prevented by adding 'Tingles' over the gaps.

Gribble infestation is concentrated on the surface of planks, and particularly at half-lap joints. This leads to loss of timber section, and the plank becomes loose in the bolted joint. Constant wave-induced motion works the plank on the bolt, enlarging the bolt holes into slots, allowing the plank to settle onto the one below, grinding away the plank edges. Eventually planks fall off, as the slots widen and the washers wear the timber.

Repairs require the removal of all loose, abraded, or infested timber, and replanking with new timber.

Half-lap joints have now been abandoned for simple butt joints, because they:

- are less prone to gribble attack

> **Box 7.2.   Durability of groynes at Bournemouth** (continued)
>
> •   are quicker to build
> •   save timber.
>
> A rolling programme of construction, now spread over 32 years, has led to steady evolution of the groyne design, improving buildability, durability and performance.

7.3.2. *Durability of connections and fastenings*

In selecting the position of fixings and connections, consideration should be given to ease of installation and removal and, in the case of bolts, also to protection of the head and nut. One method of improving protection and appearance is to use the bolt in a countersunk hole so that the head or nut does not protrude above the surface of the timber member. The head or nut is then surrounded with mastic to fill the hole and prevent the accumulation of water.

Where bolts pass through the timber they should be a tight fit. Holes should only be drilled from one side of the timber member and, where possible, nuts should be located on the more sheltered side of the connection.

7.3.3. *Corrosion of fixings and degradation of wood by corrosion products*

Certain metals can corrode in contact with water. Moreover, the breakdown products of this corrosion can have a deleterious effect on certain species of timber. Wet timber not only causes metals to corrode because wood is slightly acidic, but when a metal fixing is embedded in wood, certain conditions are created that can accelerate the corrosion of the metal. Corrosion products can cause slow deterioration of the surrounding timber, although the susceptibility of the timber is species dependent. Corrosion of the fixings and strength loss of the surrounding timber caused by corrosion can degrade joints and cause a loss in structural integrity.

Electrolytic corrosion theory can be used to explain why certain metals corrode in wet timber and why certain corrosion products are formed. For fixings embedded in wet wood, corrosion can be explained in terms of crevice corrosion. The exposed surface of the fixing in wet timber quickly displays evidence of hydroxyl ion (-OH$^-$) formation which indicates that the exposed head of the fixing forms the cathode. Therefore, the shank forms the anode of a galvanic corrosion cell. If the fixing is ferrous, for example, ferrous ions are released at the anode which promote chemical reactions causing strength loss to the surrounding timber. These reactions are triggered by soluble chlorides, present in sea water, which react with the ferrous ions formed at the anode. This results in acidic conditions forming around the shank which accelerates corrosion of both fixing and surrounding timber. Crevice corrosion usually requires an incubation period to develop, but once started it proceeds at an ever increasing rate.

Copper and copper alloys (bronze and brass) are also vulnerable to crevice corrosion although the chemical reactions differ slightly from the reactions that affect steel and aluminium-based metals. Dissimilar metals in contact in a corrosive or conductive solution, such as sea water for example, can form a galvanic cell that results in more rapid corrosion of the less corrosion resistant metal. The less resistant metal forms the anode. Understanding the galvanic series of metals allows the designer to predict the possible galvanic relationships although it is preferable to avoid designing with dissimilar metals.

Crevice corrosion and its effect on the longevity of a structure should be considered. Often the head of the fixing appears to be in good condition because it is the head which forms the cathode. Corrosion occurs at the anode and although the head appears to be sound, the shank of the fixing can have suffered severe loss of section. During inspection and maintenance works, samples of fixings may be removed and examined for crevice corrosion. Once removed, the holes should be plugged with timber and new fixings introduced into the structure.

7.4. STOCKPILING, HANDLING AND SELECTING TIMBER

Timber components must be handled and stockpiled in a safe manner and care should be taken to prevent damage. The responsibility for the timber procurement may be separated from the main contract and even if included is usually subcontracted to a specialist timber merchant. An area is usually nominated for stockpiling and the preparation of timber. This should be sufficiently large to allow for safe and efficient working and should be secure to avoid theft or vandalism of the materials.

7.4.1. Stockpiling

Good stockpiling procedure is required to prevent deterioration of timber between delivery to site and incorporation in the works, which otherwise can lead to expensive and unnecessary waste. Essential points are:

- Each type and section should have its own separate and clearly marked stockpile. Worked timber (such as shaped and shod piles) should have its own stockpile.
- Salvaged timber should have its own set of stockpiles, with timber approved for reuse separate from un-inspected, un-reworked, or reject timber.
- All timber should be stockpiled on bearers to keep it clear of the ground and to promote airflow. Bearers should be close enough to carry the section of timber in the stack without warping.
- Timber prone to rot or woodworm attack should be stacked well away from the woodworking area or salvaged material which may be affected.
- Timber should not normally be stacked more than 1000 mm high.
- Stockpiles should be protected against the weather. Even timber scheduled to be fixed into a wet environment can warp without the stability afforded by the fixings.

7.4.2. Handling

The penalties for bad handling range from breakage and premature failure of the structure to the injury or death of operatives. Clearly none of these are desirable and all are avoidable. Good handling practice includes the following:

- Bundles and wrapping should not be undone until the timber is about to be used.
- Care should be taken not to overstress the timber. Lifting at one or two closely spaced points can cause stress damage to the fibres of the timber which is hard to detect visually, but seriously weakens it, leading to premature failure in service. Forklifts should be used with particular care, as most site machines have their forks more closely spaced than timber yard ones, and they often bounce across an uneven surface, which exacerbates the problem.
- Avoidance of chain slings that are liable to damage the edges of the timber, as these cause splinters which are a health hazard. In extreme cases they can cause high point loadings, leading to similar stress damage as noted above.
- Use fabric slings with care. They should be double or choked as they are liable to slip, particularly in wet weather.
- Use control lines fixed to the ends of the load to minimise swinging, which is what usually leads to slippage in slings. The slings should be long enough to ensure that the operators can control the load without danger even if the load does get out of control.
- When using a crane to handle timber always ensure that the machine is standing on a firm level surface and never track a loaded crane.

7.4.3. Inspection

The first operation following delivery of timber to site is to check the delivery ticket to make sure the delivery vehicle has come to the right site with the right load. If certification documents are required with each load, check that these are correct and that the load markings correspond to the certification document. If the certification documents are already on site, check the load against them. Certified timber without the correct documentation and markings should be rejected. The inspection process is critical to the effective management of timber on site and adequate records should be kept, particularly if the timber is supplied by the client.

Check the load for signs of poor handling such as splinters, bruising or breakage. Compliance with the specification and/or grading should also be checked – the identification of splits/shakes, sapwood and twist or warp are usually significant. It is often appropriate to inspect a random sample of the load as soon as possible after offloading particularly if the material has been shipped from an overseas sawmill with no subsequent checking or verification.

It is good practice to agree an inspection and checking procedure at the start of a contract to ensure that there is no misunderstanding between the various parties. Suppliers will normally replace timbers demonstrated to be substandard, but the cost of offloading, inspection, storage and subsequent loading for return can cause disputes between the timber supplier and the main contractor.

7.5. CONSTRUCTION

Working near water, particularly on the coast, is seldom easy and complications including limitations on physical access, possible working times and variable ground properties, water levels, storms and loadings can all combine to make construction in coastal and fluvial areas very difficult. The use of timber does have, however, a number of advantages in that it is relatively easily to work with and can be prefabricated to a greater or lesser extent depending on the application. It remains important, however, to ensure that the structures are as simple and easy to build as possible. Key issues are discussed below.

7.5.1. Risks

Many of the most significant risks associated with timber construction on rivers and the coast are described in existing references such as *Construction risk in coastal engineering* (Simm and Cruickshank, 1998) and *Construction risk in river and estuary engineering* (Morris and Simm, 2000). Specific issues for timber structures include the following:

- Tolerances and quality of workmanship. Dimensional tolerances for timber members and constructional tolerances for installing members must be achievable. Overly severe tolerances will not contribute to the integrity or functioning of the structure and may result in significant inefficiencies in construction. In many structures different construction tolerances will be appropriate to reflect the various functions and working conditions within the structure. The simplification and/or prefabrication of some elements of the structure, such as decking panels, can prove to be particularly beneficial since it improves the overall efficiency of the construction process, facilitates improved tolerances and reduces the risk to construction operatives during in situ working. The use of alternative structural configurations should also be considered where tolerances are critical, for example planted posts may be used instead of driven piles for access steps where closer positioning tolerances are required. Similarly, the detailing of joints and connections should be carefully considered; the need for long slender bolt holes where there is a risk of deflection of the drill bit should be minimised.

- Procurement of timber. If the procurement of timber is included in the main construction contract sufficient time must be allowed in the tender period for an appropriate source of timber to be identified and in the contract programme for (at least some of) the timber to be procured and shipped prior to the commencement of the main works on site. In the case of tropical hardwoods there is often also a currency risk since the cost of the timber will often be denominated in a local currency or US dollars. The requirements for timber quality and certification also need to be clearly stated, and inspection arrangements agreed, to ensure that there is no confusion. Many of these risks can be borne more easily by the client than contractor and direct procurement of timber is often appropriate.

- Wastage and timber quantities. Contractors will need to allow for waste (including cutting, squaring and dimensional waste) when tendering for a project. This is simplified if information on all elements of the structure (including, for example, a schedule of plank sizes for a groyne) is provided by the designers. Short contract periods can significantly increase the proportion of wastage as the availability of 'stock timber' is limited and the particular section size and length required may not be readily available.

- Use of chemicals and preservatives. Particular care is required in the use of chemicals and preservative treatments, both to ensure that the site does not become contaminated and that the operative is adequately protected. Surplus materials should be disposed of in accordance with the manufacturer's guidance.

7.5.2. Plant, tools and techniques

The construction of timber coastal and river structures requires the use of a range of different types of plant and equipment, much of which needs to be operated with care to ensure that it is safe and efficient. Specific considerations are provided below.

- Plant. Many contractors used adapted and/or oversized plant for fluvial and particularly coastal works, in recognition of the greater reliability and rates of production that this can afford. Plant should be regularly maintained and operated in accordance with safe working practices.

- Handling timber. Suitable information and appropriate Personal Protective Equipment should be provided to all operatives coming into contact with timber. When tropical hardwoods are being used this will usually include a briefing on the dangers of splinters and use of good quality working gloves and overalls to reduce this danger. The density of many tropical hardwoods is such that it is only possible for very small sections to be lifted manually, and appropriate plant and equipment should be provided to allow for safe handling of timber including any prefabricated components (see Section 7.4.2).

- Air tools such as augers and wrenches driven by compressors are often used in coastal and river construction projects. These are often safer and more reliable than electrical alternatives in the hostile environment, although the use of hydraulic tools driven by diesel power packs may also be appropriate. Regular maintenance (on at least a weekly basis) is required for both the tools and the compressor. Attention should also be paid to the air hoses and a vessel to collect moisture should be provided on the compressor. Oil or antifreeze can also be introduced into the airlines to reduce moisture and prevent freezing in cold weather.

- Chainsaws. Cutting and shaping of timber is often undertaken using chainsaws, which are available in a number of sizes and with a variety of chains. Care should be taken to ensure that the saw used is appropriate to the job. Chains will require regular sharpening, particularly when used with tropical hardwoods and all operatives should be adequately trained and supervised in the use and care of the saw. Full Personal Protective Equipment, including a helmet with a full mesh mask, ear defenders, chain mail leggings, bib and gauntlets as well as steel lined safety boots will be needed by operatives using chainsaws.

- Hand tools. Tools such as the adze have been used for hundreds of years, and are still widely used for shaping of timber, including the 'topping and tailing' of piles in preparation for rings and shoes. The use of an adze requires considerable skill and training/supervision of novices by experienced operatives is essential. Suitable personal protective equipment including eye, hand and foot protection should be worn.

- Construction techniques. Consideration should be given to the weight and safe operation of all tools when developing the construction methodology. Many tools are heavy (for example a large auger might weigh 13 kg) and unwieldy to operate *(note the length of the auger required to drill through two pieces of timber in Figure 7.10)* which will reduce productivity during construction. Opportunities for full or partial prefabrication should be identified where possible as this can provide significant savings. Provision of safe access to the structure should also be considered – scaffolding is seldom practical in coastal/river locations, but bunds of beach or other materials may be used successfully. Appropriate clothing and/or lifejacket should be provided to operatives working in or close to the water: care should be taken to ensure that these are not so bulky as to increase the danger or cause a reluctance to wear them.

Figure 7.10.   Fitting ties to a pile for a timber groyne (courtesy Mackley Construction)

7.5.3. Maintenance

The maintenance of timber structures is often critical to their long term performance, and care should be taken to ensure that maintenance operations can be undertaken easily and safely. Particular issues to consider with respect to the maintenance of timber structures include:

- Members, such as cladding or planking which may need replacing within the lifespan of the structure, should be relatively easy to access and disconnect (fixings should be accessible and the need to remove other members should be avoided).

- Allowance should be made where reasonably possible for adding new primary and secondary members within the scope of the structure at a later date.
- Safe access to the structure for maintenance should be considered during the design and construction process, and measures incorporated into the structure where appropriate.
- Consistent sizes of materials and fixings should be used as far as possible to limit the range of materials held in stock and tools, etc. used on site.
- Regular monitoring should be undertaken (see Chapter 9) to enable the need for maintenance works to be identified as early as possible and minimise deterioration.

7.6. PILES AND FOUNDATIONS

Piles frequently form the primary members of many coastal and timber structures, and as such they are essential for structural integrity and the most difficult to replace if damaged or broken. It is essential therefore that the correct material is selected for the pile and that it is not damaged during construction. Particular attention should be paid to detailing to ensure appropriate durability, and monitoring should be undertaken to ensure there is no excessive wear. Pile caps should be used where possible to prevent the top of the pile deteriorating.

7.6.1. Timber as a pile material

The advantages of timber as a pile material are that it is generally less expensive than other materials and that it is more easily trimmed to length and is therefore more readily adapted to variable ground conditions. It is considered to be more environmentally 'friendly' being a natural material. In certain circumstances timber piles have a working life equal or better than steel or concrete (e.g. when subject to shingle abrasion). In general, the durability of the embedded pile is good below groundwater level but above this level timber may be exposed to biological attack by fungi and/or marine borers, depending on service conditions.

The lightweight, flexibility and handling characteristics of timber piles are suited to the construction of groynes, river walls and small jetties and walkways, providing adequate preservative treatment is given to timber species vulnerable to decay.

Timber selected for piles should be free from defects, which may affect their strength and durability. It is important that piles have a straight grain particularly where hard driving is anticipated. The centreline of a sawn pile should not deviate by more than 25 mm throughout its length, however for a round pile a deviation of up to 25 mm on a 6 m chord may be permitted.

In the United Kingdom greenheart is the most commonly used hardwood timber for piles, while Douglas fir and pitch pine are commonly used for softwood piles.

7.6.2. Pile design considerations

The working stresses expected to be encountered, both in use and during pile installation, should be calculated and checked to ensure that the permissible stress as found from BS 5268-2: 2002 is not exceeded for compression parallel to the grain for the species and grade of timber. Any loss of section due to drilling or notching must be allowed for in the calculation. Very much larger stresses are normally developed during pile driving.

There is a tendency for timber fibres to separate (broom) at both the head and toe of the pile during heavy driving. Cracking results, causing loss of structural strength and reducing the effect of preservative treatment.

Piles should be ordered in sufficiently long lengths that, after driving, they can be cut off square at the required level with the top left clean and undamaged. The squared head should be cut off to sound timber and treated with preservative where appropriate before capping.

It is recommended that where untreated softwood piles are used for the permanent foundation of a structure they should be cut off below the lowest anticipated groundwater level. They should then be extended if necessary using more durable material such as concrete. Where concrete caps are used for extending the piles they should be embedded for a sufficient depth to ensure transmission of the load. Concrete should be reinforced and at least 150 mm thick around the piles.

7.6.3. Driving procedure

Timber piles absorb energy and require larger hammers than equivalent concrete or steel piles; they are relatively easy to handle and can be driven either by using simple trestle frames and drop hammers or with more sophisticated equipment. There is however the danger that the pile head will broom and the pile toe crush under heavy driving. The likelihood of these problems occurring can be reduced by limiting the drop and number of blows of the hammer. A driving cap or 'dolly' should be used when driving. In addition, pile heads can be tapered and steel rings and shoes fitted to reduce damage. Shoes should be concentric and firmly attached to the end of the pile, the contact area between shoe and timber being sufficient to avoid overstressing during the driving operation (Figure 7.11). It is recommended in BS 8004: 1986 *Code of practice for foundations*, that it is desirable that the weight of the hammer should be equal to the weight of the pile for hard driving conditions and not less than half the weight of the pile for easy driving. The working loads on timber piles are often limited to about 300 kN on a 300 mm × 300 mm pile to reduce the need for excessively hard driving. Unnecessary high pile driving stresses can be avoided by ensuring that the pile head is central with the hammer and normal to the length of the pile. A pile should not be allowed to run out of position relative to the leader.

Piling equipment should incorporate robust gates or leaders to ensure accurate pile positioning and to eliminate twist when driving. Tolerances need to be tight perpendicular to the centre line of the groynes or similar structures to ensure a tight bearing surface for the planks and walings. They may be relaxed somewhat parallel to the centre line unless there are fixings in that plane. Suggested tolerances are 10 mm and 100 mm respectively.

Permanent tapered
mild steel circular ring *

Pile trimmed to same
taper as ring

D

D

Fixing nails

Straps

D

Diamond point
cast iron shoe

D/2

* Note: Parallel rings may be used, but are
less likely to achieve a close fit and
the pile is more likely to split

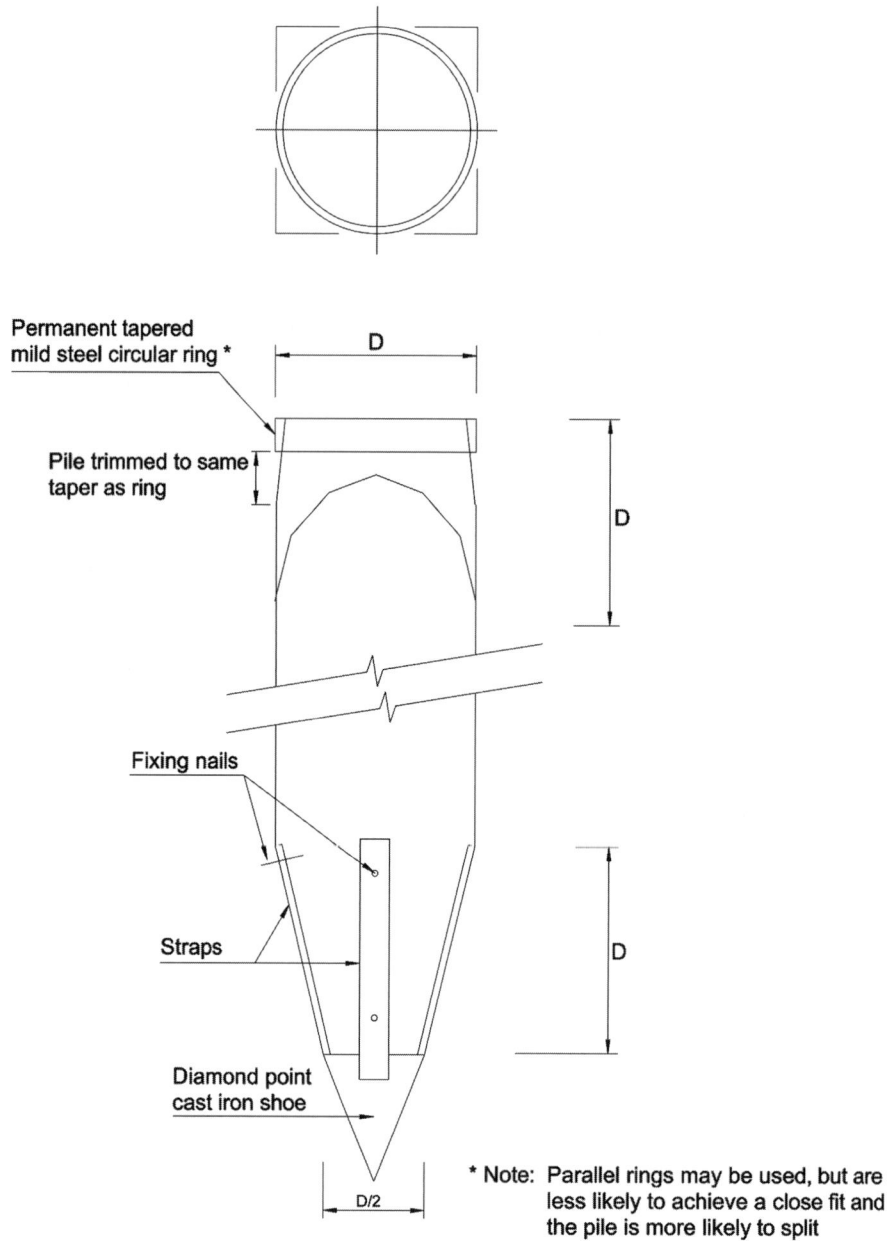

Figure 7.11.   Pile ring and shoe (plan and elevation)

Timber piles are easily shortened by cutting and if absolutely necessary can be spliced to increase their length. The tensile strength of the splices is generally low and it is necessary to ensure that the spliced ends of the timbers are squared off. However, splicing of piles should normally be avoided and Figure 6.18 illustrates measures adopted in a timber groyne to reduce the load on a spliced pile.

8. *Structure specific guidance*

The detailed design and quality of construction of timber structures can have a significant impact on their performance and durability, but these are frequently only appropriate on a limited range of structures. Thus, whilst the general principles and issues have been described in the previous two chapters, the following sections deal with those of specific relevance to the various different structures. In each case the relevant issues are discussed in the following order:

- choice of timber structure
- operating conditions
- timber species
- overall configuration
- overall stability
- members and connections
- details
- finishes and fittings
- maintenance.

References to relevant literature or other sources of information are also provided where available. It should be remembered, however, that the design of any structure must include consideration and analysis of its particular functional requirements, situation and conditions. This analysis must be undertaken by competent individuals or organisations and the experience available from specialist consulting engineers is often valuable.

8.1. GROYNES

Traditionally, timber groynes in the UK have been impermeable vertical structures constructed more or less perpendicular to the shore- or bank-line to restrict the movement of sediment. They are used on the open coast to retain beaches and within rivers and estuaries to control the flow and sedimentation regimes. Conventional groynes usually work by fulfilling one or both of the following functions:

- Providing a physical barrier to the movement of the beach material, enabling the beach within the groyne bay to reorientate approximately perpendicular to the incident wave direction.

- Reducing or diverting currents away from the beach or river bank.

It must be noted that groynes do not prevent cross-shore transport of sediment on beaches and often only interrupt the longshore drift, transferring the erosion problem down-drift of the groyne field if designed in isolation. Coastal groynes do not usually contribute to the primary function of coast protection or sea defence schemes (the reduction in wave energy reaching the backshore area), but retain and sometimes build-up the beach, which thus provides protection by absorbing part of the incident wave energy.

A variety of permeable timber groynes has also been used. These range from round logs placed vertically, with the spacing varied to alter permeability, to rather more complex structures constructed from sawn timber, with either a straight, or zigzag plan shape.

8.1.1. Choice of timber structure

The other main construction materials for groynes are rock, masonry and concrete. Steel may also be a significant component when used in conjunction with other materials. The choice of material for a particular situation depends upon technical, environmental and economic considerations.

Timber is probably the most technically versatile of the materials. It has the potential to provide the best balance in terms of strength, size, durability, workability and buildability. At the design and construction stages it is easily adapted to the situation and during the service life of the structure it lends itself to modifications (e.g. raising or lowering of planking heights) and to maintenance (e.g. replacement of defective planks). Whilst other materials may perform better in certain areas (e.g. rock is more durable) they are likely to be less effective in others.

Environmentally, timber has a number of benefits. A timber groyne usually takes up less space than its equivalent in other materials and in an amenity situation is often more compatible with its surroundings. Where space and aesthetics are less critical, other materials may have the advantage.

Economically, timber structures may have lower capital but higher maintenance costs, which can include periodic replacement. Comparison with other materials has to be made on a whole life basis and durability is often a key factor in this consideration.

8.1.2. Operating conditions

Timber groynes can be used in most coastal locations where it is considered feasible to retain a beach, and there are many examples of the successful application of groynes on different types (including both sand and shingle) of beaches. Rock structures, or heads, may be preferred on beaches with a high wave energy as they are less reflective and reduce scour at the tips of the groynes. Similarly, although timber piles can be excavated and planted into hard substrate, other forms of construction may be more practicable in these situations.

Coastal groynes will normally be subject to considerable abrasion from mobile beach material and are often also exposed to marine borers below the mean water level. They should be designed to resist wave loading and appropriate differential beach levels across the structure. Safety of beach users will need to be considered during the design and detailing of the structure, and groyne markers are often required to assist both small craft and swimmers.

The operating conditions for river groynes will usually be much less onerous than those for coastal groynes, although estuarine locations may still be subject to some wave loading and/or marine borer attack. The provision of adequate navigation marks is particularly important to prevent damage to or from river groynes.

8.1.3. Timber species

The requirements for strength and resistance against both abrasion and marine borers limits the selection of timber species for many coastal groynes to tropical hardwoods, most commonly greenheart or ekki, however other species such as balau and kapur have also been used with some success. Temperate hardwoods such as oak and softwoods such as pitch pine or Douglas fir have also been used, but are less durable and more susceptible (unless preservative-treated) to attack by marine borers.

The selection of appropriate timber will be dependent on environmental constraints as well as the exposure and required working life for the structure. In some cases, different species may be used for different components (e.g. tropical hardwood piles and walings and softwood planks) or exposures (above/below normal beach or water level) but this can make procurement difficult and complicate construction. Construction of groyne fields may require particularly large quantities of materials, necessitating a significant procurement period and good co-ordination to match the timber supply to the progress of the works.

8.1.4. Overall configuration

The overall configuration of a particular groyne or groyne field (Figure 8.1) is dependent on the characteristics of the beach (including the type and grading of the beach material) as well as the wave, wind and current exposure which will all affect the beach profile and rate of transport. Guidance on the design profile of groynes is available from a number of sources including CIRIA (1996), Fleming (1990) and USACE (1992). Most schemes will be optimised using physical and/or numerical models. However, it should be noted that the models and expected loadings include inherent uncertainties, and provision within the design to vary the profile by adding or removing one or more planks can be very valuable.

Some of the different options relating to the configuration of timber groynes are illustrated in the following sections.

Conventional groynes

Conventional impermeable groynes are constructed from sawn timbers including square vertical piles, planking (normally on the up-drift side) and often one or more horizontal or sloping walings to distribute loading between the piles. Planking or timber sheet piles are often continued some distance below the beach level to provide passive resistance against ground and wave forces. In some cases, props or ties (see Section 6.5) will be needed to provide stability, although they can suffer from extensive abrasion and the use of double piles may also be considered. Conventional impermeable groynes reflect waves, and may focus longshore currents at the tip of the groyne, both of which can lead to scour.

Figure 8.1. Conventional impermeable timber groyne field (courtesy Arun District Council)

Permeable groynes

Permeable groynes are designed to allow beach material to pass through them but at a slower rate than would otherwise be the case for an open beach. In practice it is very difficult to achieve this, especially with narrow structures. The driving forces for the longshore movement of beach material can often dominate the structure, resulting in very little net effect upon overall beach levels: also, the passage of beach material through the structure can cause severe abrasion of the timbers.

The use of permeable groynes, instead of more conventional impermeable groynes, is usually with a view to reducing their impact upon longshore movement of beach material and the potential for realignment of the coast down-drift of the groynes. The use of lower and/or shorter conventional groynes can have the same end result. However, with careful design and in selected situations, including those with lower energy regimes, permeable groynes have a role to play. Additionally, such structures can be combined with other concepts, such as zigzag shapes, to improve their performance.

Permeable groynes may be used on beaches, but are widely perceived to function best in locations where there is a current but minimal waves, particularly rivers. They often consist of single or double rows of round timber piles driven into the underlying strata (Figure 8.2). The use of round piles minimises wastage but it should be remembered that such materials will be susceptible to marine borers if temperate timber is used or sapwood is not removed. The spacing between the piles and level of the piles above the bed or beach will depend on the intended influence of the groyne, but whilst the spacing is often approximately equal to the diameter of the pile, no established relationships are available for design purposes.

Permeable groynes constructed from sawn timbers have been used at a number of locations around the British coast but anecdotal evidence suggests that their effectiveness and durability is poor compared to conventional impermeable structures.

Figure 8.2. Pile groyne at Ameland, The Netherlands (courtesy HR Wallingford)

Zigzag-shaped groyne

The main purpose of a zigzag shape is to prevent waves or currents running parallel with the structure and causing scour. The shape also increases the effective width of the groyne which can further reduce the possibility of down-drift scour whilst providing the structure with inherent stability; detailing can be more difficult.

When a zigzag configuration is combined with a permeable construction, the beach material partially trapped within the zigzags can act as a further filter medium in conjunction with the structure itself.

Geometry

There is a range of different groyne concepts, which provide appropriate solutions in particular locations. One example, illustrated in Figure 8.3, uses a combination of recycled and local timber to construct zigzag shaped groynes. Although these structures have a limited life and high maintenance commitments, their low capital cost makes them an attractive option in some situations. If new tropical hardwoods are used, initial costs will be higher, but they should require less maintenance.

Figure 8.3. Low cost groyne, Calshot, Hampshire (courtesy HR Wallingford)

A variety of other types of groyne has also been used. For example, plank pile groynes consisting of two plank pile walls with ties between the piles and the space between them filled with sand, gravel, earth or stone. These structures may be used as small breakwaters and where they are overtopped rubble is placed as a top layer to prevent the finer material washing out. Other examples include the use of a single tongue and groove plank pile wall supported by sawn timber walings and round piles. The form and geometry of the groyne will need to be optimised for the material used and the loadings occurring at the particular location.

Overall stability of structure

The overall stability of the structure will need to consider wave loadings and large differences in beach levels across the structure. The techniques described in Section 6.5 are generally appropriate for groynes and the technique adopted for any particular structure is largely a function of the ground conditions at the location.

Members and connections

Since groynes are heavily loaded and often constructed in difficult working conditions, the key to successful design of members and connections is keeping the structure arrangement simple with as few connections as possible. Prefabrication of elements such as panels of sheet piles, fixed near or below low water, can increase safety and improve the quality of construction. Much useful information on members and connections is given in Sections 6.5 and 6.6.

Preservative-treated timbers are not usually used in groynes due to the high propensity for abrasion, which could quickly strip the preservative protected outer layer of the timber leaving the remaining timber exposed to biological attack as well.

Finishes and fittings

All fixings on groynes (see Box 8.1) are exposed to severe wear and oxidation, due to the aggressive marine environment. Therefore as few bolts as possible should be used and never placed in the same grain line of the timber. Galvanised steel fixings are commonly used but stainless steel fixings have the advantage that the groyne can be dismantled or refurbished more easily and the fixings can be reused.

As with the arrangement of members and connections, finishes and fittings should be kept as simple as possible and as few as possible.

Box 8.1. Groyne fixings

Bournemouth Borough Council started building timber groynes in 1971. Aluminium bronze bolts were used at first but large numbers were stolen because of their high scrap value. Stainless steel became the preferred option in 1973, because of its resistance to corrosion and abrasion, and its relatively low scrap value. Lock nuts were essential to stop the nuts from loosening, but some managed to anyway.

Around 5% of the new bolts were unusable because of thread damage, presumed to have been incurred during transport and handling. Furthermore, grit in the threads caused nuts to lock up, giving additional wastage. However, when groynes were rebuilt after 25 years service, the nuts could often be undone, with potential for reuse.

Drilling holes for long bolts is a critical operation as 1mm clearance only is desirable to limit component sag, so good quality bits are essential. Long drill bits are flexible and can wander, which means that a sledgehammer is needed to drive the bolts home. This adds to fixing time, and is not good for the bolt.

Risk assessments indicated that protruding bolt threads were undesirable so, in 1999, coach screws were considered. As part of the assessment of the potential, the current contractor (Dean & Dyball) tried using tungsten carbide tipped bits to get maximum accuracy when hole drilling, and conducted pull-out tests using new greenheart baulks. These were impressive. The stainless steel coach screws failed by stretching before the threads pulled out. As a result, all subsequent groynes were built using coach screws wherever possible.

Three years' experience shows that coach screws have the following advantages over bolts:

- they are cheaper
- they are not susceptible to thread damage
- they can be fixed from one side of the groyne only, reducing excavation and labour requirements
- they are less of a danger to beach users
- they can be done up tighter than bolts
- hole length, and hence drilling time, is reduced
- there is no risk of the drill bit going right through the timber and producing splinters.

Maintenance

Maintenance is critical to the effective performance of timber groynes and can range from removing the top plank if beach levels have fallen (to reduce wave reflections from the structure) to replacing life-expired piles or increasing the level of planking on a groyne by adding pile extensions.

Abrasion

To protect piles from abrasion as shown in Figure 8.4a, softwood or recycled timber rubbing pieces can be attached to the piles at critical levels (Figure 8.4b). Extending planks beyond piles may also reduce wear, increasing the life of the primary and secondary members. Members should be sized with an allowance for wear and the connections carefully chosen. On some sand beaches ply panels have been fixed to the groynes preventing sand being transported between the planks and thus minimising abrasion, which increases dramatically with the size of the gap. Maintenance will require regular inspection of the structures and replacing of worn rubbing pieces.

Adaptability

Groynes should be adjusted to the optimal height, matched to the changing beach profile, possibly several times throughout their lifetime. In some cases, extending part or all of a groyne to the preferred height may be an attractive solution instead of rebuilding an entire groyne. However, when constructing a groyne, an allowance may be provided in the pile length for extra planking. See Figure 8.5 and also note the 'shaped' planks at the interface of the old and new sections of planking which were used to adjust the profile of planking to match the extended piles.

Figure 8.4a. Abraded timber groyne pile (courtesy HR Wallingford)

Figure 8.4b.   Rubbing pieces protecting the pile on a timber groyne (courtesy HR Wallingford)

Figure 8.5. Pile extensions to facilitate reprofiling of a groyne field (courtesy Arun District Council)

Reprofiling can be carried out gradually, i.e. a few planks at a time to encourage natural accretion, or in one operation in conjunction with beach nourishment.

Construction

As groynes are generally built using land-based plant, special attention should be paid to constructing the seaward end of the groyne as the tidal window may be very limited.

Literature

CIRIA (1996). Beach management manual. Simm, J.D., Brampton, A.H., Beech, N.W., and Brooke, J.S. Construction Industry Research and Development Association, Report 153.

Fleming, C.A. (1990). Guide on the uses of groynes in coastal engineering. Construction Industry Research and Information Association, Report 119, London.

USACE (1992). Coastal groins and nearshore breakwaters. US Army Corps of Engineers, Engineer Manual 1110-2-1617, Aug.

8.2. DUNE PROTECTION, BREASTWORKS AND REVETMENTS

8.2.1. Dune protection

Several techniques have traditionally been used to protect, stabilise or restore dunes, often using timber or brushwood cuttings. It should be remembered that vegetated sand is the best protection, and the aim of stabilising dunes. Dune or sand fences (Figure 8.6), placed parallel to the shore reduce wind speed across the sand surface and thus encourage the deposition of wind-blown sand. Further the fences will help protect existing or planted vegetation by stabilising the surface and reducing trampling. Another use of sand fencing is to keep roads or paths leading through the dunes free of drifting sand.

Timber fences may be constructed as barrier fencing at the base of a dune (Figure 8.7), protecting it from erosion due to occasional wave attack. Sand moves onshore with low waves and offshore when high waves attack the shore during storms. Protecting dunes against storm waves is an effective method to decrease erosion. The most common type of wave barrier fencing is known as a Dutch fence, constructed of brushwood or chestnut paling in a rectangular shape at the toe of the dune. More solid structures can be built for the same purpose, including vertical or sloped permeable revetments. Wave barrier fencing should always be built above the limit of normal wave run-up to avoid undermining.

Laying a thatch of brushwood over the surface of the sand and planting marram grass or sowing grass seeds in the gaps can help recover areas with no or little supply of sand. If willow and poplar species are used, the thatching can be dug into the sand to encourage growth from the cut stem.

Dunes are often stabilised or restored in a management scheme of dune grass planting, thatching, fencing and beach nourishment.

Figure 8.6. Dune fencing, Greatstone on Sea, Kent (courtesy HR Wallingford)

Figure 8.7. Dune barriers at Pwllheli, Gwynedd (courtesy Posford Haskoning)

Choice of timber structure

As natural materials, timber and brushwood are more appreciated aesthetically than synthetic materials. Sand fencing can best be constructed using biodegradable materials, as the fences often will become buried if successful or be washed away if not. For the same reason it is preferable to use low-cost materials such as brushwood. However, fencing is usually constructed using long lasting wire which may become a hazard in the longer term. Although natural materials are preferred, synthetic snow fencing has successfully been used for dune fencing and its use may be considered in some cases.

The labour costs for dune management schemes can be considerable although, in many cases, conservation volunteers may carry out the necessary maintenance, as the dunes are often of high ecological value. Straw bales are successfully used to repair small blowouts.

Operating conditions

Dune fencing, wave barrier fencing and thatching will resist some erosion, but cannot prevent erosion where wave attack is continuous and damaging. Where wave attack does occur regularly, it may be necessary to place substantial timber piles to avoid annual reconstruction. To prevent rotational failure support piles must be buried adequately, as beach levels will drop during storms.

Wave barrier fencing, placed just in front of a dune helps prevent severe damage caused by storm wave attack. The rectangular-shaped Dutch fences have two lines of brushwood, the outer acting as the main barrier against waves and the inner protecting the dune base. Both dune fencing and wave barrier fencing may be removed for part of the year on high amenity value beaches. However, this method is not recommended as the extra work may damage the dunes and the fences may not be beneficial at all in the end. To reduce the amount of work on temporary fences, permanent posts can be erected, to which the fencing is attached seasonally. A sloped timber revetment as a wave barrier fence will suffer less toe scour than a vertical wall, but cannot resist continuous wave attack either.

Public access routes must be built in to the design of dune retaining structures, especially on popular public beaches to prevent the public from damaging fences to create their own paths. The tracks should follow the natural contours of the dunes as much as possible. In popular areas fences should be designed to be as vandalism-resistant as possible, e.g. with secure fixings. Dune fencing adjacent to paths will stabilise the sand and prevent trampling of the dunes.

Timber species

The choice of materials for dune fencing or thatching will depend on required life, length of the frontage, level of maintenance and potential for vandalism. Beach users often remove thatching and fencing materials for bonfires. Fencing materials include chestnut palings, pine, birch, wooden slats or synthetic fabrics. Brushwood can be made-up of a number of different species, e.g. local sea buckthorn or conifer trimmings. Locally available timber offcuts can be used but may result in being hazardous, as they are likely to generate splinters and loose staples.

Sea buckthorn is likely to spread uncontrollably if the branches are not dead or still have berries on them.

Wooden slat fences are not recommended, as they are relatively expensive to construct and often are stolen by beach users. Wooden slats tend to deflect the wind, causing erosion whereas brushwood traps sand because of its filtering action.

Posts and wave barrier revetments can be constructed from treated softwoods or in some cases hardwood, possibly recycled, as the lengths required are short. Railway sleepers may be used to create vertical wave barrier fencing in more vulnerable areas, acting more like a sea wall rather than fencing.

Overall configuration

Brushwood fencing and chestnut paling are the two most common fence types. Brushwood fencing is an upright hedge of stacked brushwood whereas chestnut paling is constructed using support posts and tensioning wires.

Fencing should be aligned parallel to the dune face, forward of the toe of the dunes, although in some cases the fencing is set-up at right angles to the prevailing wind. Posts and wires are best set-up to last for several years as they are costly to replace. The posts should be long enough to ensure the stability of the fence at the lowest expected beach level. Sand will build up 1.25 m leeward of the fence (Ecopro, 1996). Initially fences should be constructed at the downward end of where the build-up is required. When the sand level has increased, additional lines of fencing can be added, spaced at a distance of four times the height of the fence (Ecopro, 1996).

Short spurs may be constructed up a dune, spaced at six times the fence height (Ecopro, 1996), at areas subject to local erosion to restore that specific part of the dune. Spurs running seaward are less useful and may be damaged by swash zone debris or by beach users trying to walk along the shore at high tide.

A porosity of 30% to 50% is required for dune fencing to be effective (Scottish Natural Heritage (SNH), 2002). Standard paling fences of the type used around construction sites have a porosity larger than 50% and should therefore not be used for dune management.

Wave barrier fencing

Constructing wave barrier fencing may generally be the same as dune fencing, other than the rectangular layout of the wave barrier. Dutch fences are constructed using brushwood and arranged in the form of a rectangle. Trenches are dug at the toe of the dune in which the brushwood is placed upright. The primary function of wave barrier fencing is to absorb wave energy, but they will encourage sand build-up too.

Dune thatching

Thatching should cover 20 to 30% of the sand surface (SNH, 2002) and be laid above the line of normal wave run-up. Biodegradable mats can be used for the same purpose but are more likely to be damaged in the marine environment. Dune thatching or matting should always be complemented with the planting of marram grass or sowing of grass seeds in the gaps or in holes. It is recommended not to use thatching on slopes steeper than 1:2. In these cases the dunes will have to be reshaped or renourished to obtain the required dune profile.

Maintenance

Regular maintenance is required for both fencing and thatching, and repairs should be carried out as soon as possible. Maintenance can be carried out during the whole year, but planting of marram grass is preferably done in spring. When planning maintenance, nesting and migration of birds, public access and safety should be taken into account. Damage on thatching should be repaired as soon as possible.

Dune fencing may catch blown litter.

Literature

Brookes, A. and Agate, E. (2000). *Sand Dunes: a practical handbook.* British Trust for Conservation Volunteers.
http://handbooks.btcv.org.uk/handbooks/index/book/137
Scottish Natural Heritage (2002). A guide to managing coastal erosion in beach/dune systems. http://www.snh.org.uk/pdfs/publics/catalogue/nhm.pdf
Searle, S.A. (1975). *The tidal threat.* Dunes Group, UK. ISBN 0950400106.

8.2.2. Breastworks and revetments

Seawalls, revetments and breastwork are generally positioned at or slightly landward of the crest of the beach aligned parallel to the shoreline. They are normally intended to provide protection against erosion or breaching of the beach during severe events and often take a similar structural form to groynes. Breastwork is often used as a secondary defence at the back of a groyned beach.

Choice of timber structure

Shore parallel structures may be constructed from a wide range of materials including steel sheet piling, rock armour and concrete. Timber has the advantage (along with steel and, to a lesser extent, concrete) that it can be used in a vertical structural form, reducing the volume of material required and minimising the area of beach occupied. Such vertical forms will result, however, in the reflection of incident wave energy and is likely to increase erosion of the beach.

Where dunes or a shingle ridge are located behind the beach the use of impermeable breastworks or seawalls will severely disrupt interchange of material to and from the beach, limiting the ability of the natural system to roll back and adapt to changes in loading.

Operating conditions

Seawalls and breastwork should be designed to withstand the full force of wave impacts and to accommodate any expected beach level changes due to erosion and/or scour. The term 'bulkheads' (Figure 8.8) is often used to refer to structures primarily intended to retain fill and generally not exposed to significant wave action. Typically, revetments are lighter, permeable structures allowing some movement of beach material through the structure.

Figure 8.8. Timber bulkhead with rock toe protection at Lepe, Hampshire (courtesy HR Wallingford)

Timber species

Conventional impermeable breastwork has traditionally been constructed from similar materials to conventional groynes – including tropical hardwoods. However, where the structure is located landward of the high water level the susceptibility to attack by marine borers will be significantly reduced (even if the water table permeates through the beach) and less durable timbers (such as oak) may be used. Temporary or low-cost structures may be constructed using round piles, or a combination of treated softwood posts and geotextiles.

The selection of material will depend to a significant extent on the likelihood of attack by marine borers and the degree to which the structure will need to resist abrasion from beach materials.

Overall configuration

Most timber breastworks/bulkheads are impermeable vertical walls, which will reflect any waves reaching the structure. Such structures are unlikely to be appropriate in areas of very high wave energy, or where there is no beach in front of the structure.

Conventional breastwork

The construction of conventional breastwork is shown in Figure 8.9 and a completed breastwork at Pevensey, East Sussex in Figure 8.10. The configuration is very similar to a conventional timber groyne, with piles driven into the beach and planking or plank piles used to provide passive resistance.

Figure 8.9. Construction of a timber breastwork (courtesy Mackley Construction)

Figure 8.10. Breastwork at Pevensey, East Sussex, retaining substantial shingle beach, to right of photo (courtesy Aitken & Howard Ltd.)

An example of a vertical permeable breastwork at Shoreham, West Sussex is shown in Figure 8.11. This solution is not very common and has the disadvantage that the individual vertical piles do not wear well.

Figure 8.11. Vertical breastwork at Shoreham, West Sussex (courtesy Shoreham Port Authority)

Permeable revetment

An example of a permeable sloping revetment is illustrated in Figure 8.12. These structures were used extensively to protect soft cliffs on the east coast of England, but are widely regarded to be of only limited effectiveness; many lengths have now been replaced by rock armour revetment. In some locations the timber framework is used to secure smaller rock armour that would otherwise move under the force of waves as in a gabion or crib work structure.

Permeable revetments reduce the reflection of wave energy, limiting scour problems at the toe of the structure, and allow the cross-shore movement of material. However, the structures may limit access to the foreshore (note the steps over the structure in Figure 8.12) and have are susceptible to damage and difficult to maintain.

Figure 8.12.   Permeable sloping 'Mobbs and English' revetment (courtesy HR Wallingford)

Low cost structures

In some locations lower cost structures have been constructed using softwood timber posts and geotextiles (see, for example, Figure 8.13). These are normally unable to withstand larger wave forces or have limited service life but may provide a reasonable standard of protection in locations with low wave exposure.

Figure 8.13.   Low cost revetment with geotextile at Calshot, Kent (courtesy HR Wallingford)

Overall stability

Scour of the foreshore due to wave reflections is the largest problem with seawalls and revetments. As scour increases, so does the water depth, allowing higher waves to reach the wall and thus eventually undermining the foundations and causing collapse. Sheet piling, together with bed protection such as matting and rock armour at the toe of the structure, is often used to protect against this failure mechanism.

Seawalls or revetments disturb the natural longshore drift of sediment and might result in extra erosion at the down-drift end of the structure. In some cases outflanking may occur, undermining the end of the wall from behind. By feathering the end of a revetment back into the dunes this problem can be prevented.

Maintenance

As with groynes, these structures have a working life of around 30 years before needing major replacement work, given regular inspection and replacement of damaged or worn parts.

8.3. WAVE SCREENS

Wave screens (Figure 8.14) are permeable fence-like structures normally used to protect moorings or marinas from excessive wave action, although they can also be used in a similar manner to nearshore breakwaters protecting a length of coastline.

Choice of timber structure

Large wave screens are frequently constructed from a range of different materials; tubular steel piles are often used for the main piles, with timber or steel walings and timber planks forming the screen itself. Smaller screens may be constructed solely from timber.

Timber species

The selection of the most appropriate timber species will be primarily dependent on the environmental conditions at the structure. Greenheart is often used for larger screens as it has the durability to provide a significant service life before replacement, however the use of preservative-treated softwoods could be considered.

Overall configuration of structure

Wave screens for moorings and marinas are usually constructed from vertical planks fixed to horizontal walings. This enables the plan shape to be more complicated than would be the case with horizontal planks. However, if there is a limited tidal range the replacement of long vertical planks due to the deterioration of a small section at the water level could be wasteful. Wave screens are often constructed on the sides of piers and/or jetties to give some shelter to small boats embarking passengers.

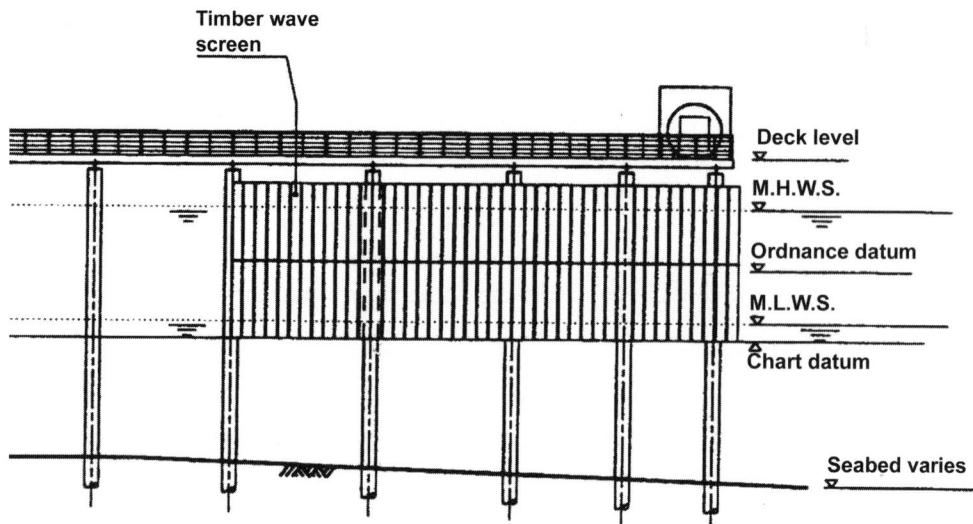

Figure 8.14.   Elevation of a wave screen (courtesy Posford Haskoning)

Screens on the active beach must resist severe wave impacts and these often have sloping faces to maintain their stability without deep foundations. On amenity beaches, screens should be designed with sufficient access possibilities but it is preferable to build no such structures at all on these beaches.

The spacing between planks can be adjusted to control the proportion of wave energy transmitted or reflected.

Wave barrier fencing can be found on the upper beaches, minimising the erosion of dunes in case of a storm event, and may be constructed using chestnut paling as described in Section 8.2.

Overall stability of structure

The overall stability of a wave screen will be similar to that of a pier or jetty, although the additional lateral load caused by the screen must also be considered.

Members and connections

Timber members should be sized to withstand the applied loadings, with an additional allowance for reduction in section due to abrasion or other deterioration. Figure 8.15 shows a plan view of part of a wave screen using 225 mm × 75 mm greenheart planks with a relatively small spacing. Note that in this instance only the planks were timber, with steel main piles and waling proving more economical whilst the timber planks still provide an attractive appearance.

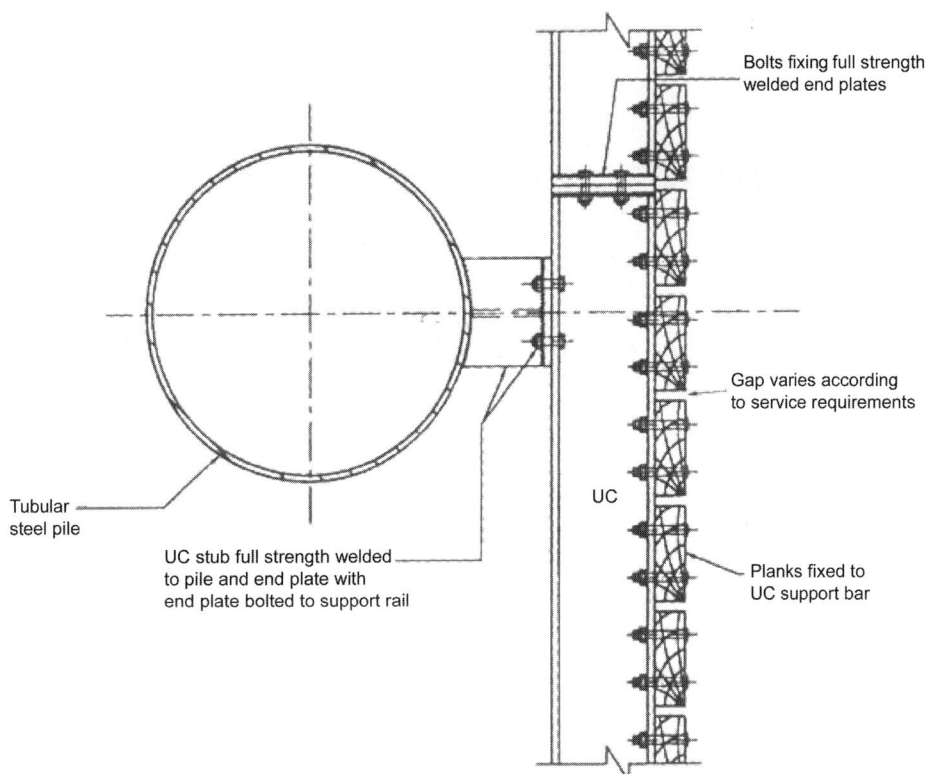

Bolts fixing full strength welded end plates

Gap varies according to service requirements

UC

Tubular steel pile

UC stub full strength welded to pile and end plate with end plate bolted to support rail

Planks fixed to UC support bar

Figure 8.15.   Plan detail of a wave screen (courtesy Posford Haskoning)

Literature

Gardner, J.D., Townend, I.H. and Fleming, C.A. (1986). The design of a slotted vertical screen breakwater. *Proc 20th Intnl. Conf. Coastal Engineering*. American Society of Civil Engineers.

McBride, M.W., Smallman, J.V. and Allsop, N.W.H. (1996). *Guidelines for the hydraulic design of harbour entrances*. HR Wallingford Report SR 430, Feb.

8.4.   JETTIES, PIERS, PONTOONS AND DECKING

Timber decking is commonly used on jetties, piers, catwalks and floating pontoons. As shown in Figure 8.16, it provides a durable and attractive finish.

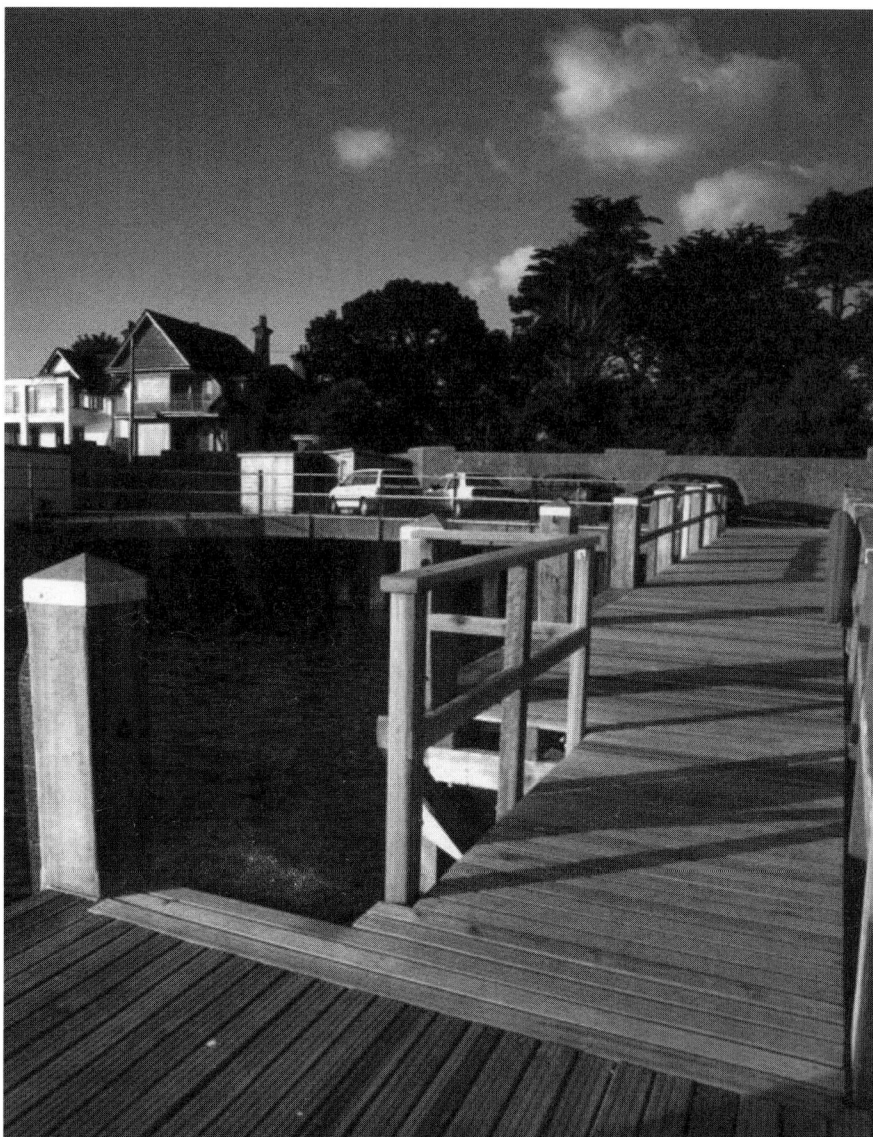

Figure 8.16.   Sandbanks Jetty (courtesy Dean & Dyball)

Choice of timber structure

The main advantages of timber are its high strength to weight ratio (on floating catwalks and piers lightweight decking requires smaller floats) and pleasing appearance. In marinas timber is an excellent material, as boats will not be damaged when they come in contact with a pier or catwalk. Some alternatives, such as recycled plastics, often leave black marks on light coloured vessel hulls that can be the cause of complaints by owners of pleasure craft.

Timber species

Timbers vary in their wear resistance to pedestrian and other traffic when employed as decking, and the use of species such as Douglas fir, Baltic redwood and dahoma is best confined to areas of light traffic. Pitch pine, dark red meranti, keruing, kapur, opepe, jarrah and karri, for example, have sufficient resistance to wear to recommend their general use for decking, except where resistance to heavy pedestrian traffic (i.e. of the order of 2000 people per day in well defined traffic lanes) or vehicle traffic, and resilience to impact loads is required, when oak, danta, belian, greenheart and okan, amongst others, would be preferred.

Design of decking

The functional requirements of a deck will vary considerably depending on such factors as the level of vehicle and pedestrian traffic, the degree of exposure, environmental conditions, or the particular activities the deck must accommodate. The information given in this section assumes that the deck will only be exposed to conditions described as Hazard class 3.

By establishing principles of good practice in design and specification the aim is to ensure that wooden decks, whether industrial or public, will not only be suitable for their purpose but will also meet the desired service life, with the minimum of maintenance. To achieve this, designers, contractors and clients need to understand the quality of design, materials and construction required in building decks. For this reason it is important to choose material suppliers and contractors who have experience in the supply of the materials for decking and their construction.

Long term durability as well as, perhaps, the appearance of wooden decks, will also depend a great deal on the correct choice of materials, good detailing and careful construction. Decking is, by its very nature, some of the thinnest timber met with in marine and fluvial construction; hence, any degradation will reduce the working section disproportionally when compared with other timbers. Because decks are fully exposed to wetting and drying cycles it is of primary importance to choose a timber that is rated 'moderately durable' or 'durable', from which the sapwood is excluded, or to use timber that has been pressure-treated with preservative to enhance its durability. If the deck is for industrial purposes or provides for vehicular access, timber with high strength properties should be specified.

Durability is a primary concern because it has not only financial but also significant safety implications. A failure is likely to have serious consequences particularly if it results in a hazard to the public. With this in mind, it is essential that the designer and builder are fully conversant with Sections 6 and 7 of this manual, and put these recommendations into practice. These principles should be applied not only to the deck itself but to any substructure and ancillary components such as handrails, steps, balusters, seats, etc.

Other points to note are:

- In some cases, deck boards are used 'undried' in order to save the cost of kiln or air drying. In this case it is essential to fix frequently along the length of the board and also to allow for any shrinkage across the board width as it dries. This situation is typical of temperate hardwoods, such as European oak, but timber cut from large imported tropical logs can also be initially at quite a high moisture content. To reduce the effects of shrinkage across the board as much as possible it is advisable to use relatively narrow boards, no wider than 150 mm and preferably less.

- Softwood boards will generally absorb moisture more readily than hardwood, but there are various treatments available to make them more moisture resistant and consequently more dimensionally stable.

- Small surface cracks or 'checks' often occur in deck boards due to drying out when exposed to hot sun. These will usually close up when the surface is wetted, but to minimise the occurrence of these 'checks' it is preferable to choose timbers that are rated as having 'small' or 'medium' movement, as there will be less risk of checks or splits developing as the surface moisture content is reduced.

- Deck boards are subject to gradual wear and erosion over a period of time depending on their use. Some care should be taken in the initial choice of a timber to stand up to the anticipated end use. As a general rule, timbers of lower density will be subject to greater wear than those of higher density.

- Wetting from seawater in the form of spray or splashing is actually beneficial in that the salt tends to inhibit the growth of fungi. While it is always advisable to use stainless or hot galvanised fixings and fittings for decks, this is particularly important in a maritime environment where there is a high risk of corrosion and consequent risk of staining to the timber.

Finishes and fittings

Non-slip surfaces can be produced by cutting grooves into the decking. However, this may not be adequate in some situations (such as where the slope of the deck is very steep or continually wet). In these situations it may be necessary to create a non-slip finish by applying an epoxy resin with grit to the decking. Some proprietary boards are available with the non-slip finish applied in factory conditions for improved durability.

Fixings should be countersunk to avoid forming a trip hazard. The hole will then need filling with a resin or wooden plug to prevent water collecting in it and damaging the timber or connection. Such activities are labour intensive (see Figure 8.17) and the use of wooden plugs can be expensive.

Small differences in the thickness of sawn decking boards will often result in an unacceptable surface and/or trip hazard. Decking is often specified as 'planed all round' to ensure that boards are all the same thickness, thereby avoiding problems with appearance and safety.

Figure 8.17. Covering decking fixings with wooden plugs (courtesy HR Wallingford)

Maintenance

Regular inspection and maintenance is required to ensure decking does not present a trip hazard.

See Box 8.2 for case study on Saltburn Pier refurbishment.

Literature

TRADA Technology (1999). *Timber decking manual.* Timber Research and Development Association, High Wycombe, UK.

BS 6349-1: 2000. *Maritime structures. Code of practice for general criteria.*

Box 8.2.   Case study, Saltburn Pier refurbishment

This pier was originally constructed in 1860 with a hardwood deck supported on cast iron columns and piles. It is the last remaining pier on the north east coast, and faces slightly east of north, making it particularly vulnerable to the notoriously changeable weather of this part of the coast.

Courtesy John Martin Construction

It had been founded on a local shale which, over 140 years, had been eroded. The works entailed were:

- deck replacement
- pier leg refurbishment
- installation of new foundations where the existing ones had been undermined.

In addition, access was difficult, especially to the pier head, where work to the support structure could only be executed on selected low tides, the deck fixing lay firmly on the critical path, so any delays met with in that operation would feed directly into and extend the programme, and there were health and safety

Box 8.2. Case study, Saltburn Pier refurbishment (continued)

implications to fixing the superstructure timbers in situ. Add to this the fact that the support structure works required the use of a heavy crane which was subject to a lot of down time waiting for suitable tides, and which was therefore virtually free issue as far as deck construction was concerned, and the situation was ripe for a proposal for prefabrication of the deck. This was accepted, and the deck redesigned to facilitate the operation. The advantages to the client and contractor were:

- the members could be assembled in workshop conditions, giving good quality and health and safety control
- risk of contract overrun was reduced
- it gave work to local industries
- it saved money.

Other points to note are:

- The timber selected for the deck was ekki. According to Table 4.5, this has a large movement value. The decking was designed for a span of about 1500 mm, but this raised the possibility of a trip hazard caused by movement. An additional non load-bearing stringer was introduced to reduce the potential for relative movement.
- Due to high production standards it was possible for the deck boards to be sawn to thickness rather than planed all round (PAR).

All fixings were made of stainless steel and the coach screws holding down the decking were countersunk. The heads were then plugged with a hard setting mastic – bitumen was considered at one stage but rejected as it was felt that it could soften during hot weather, providing a hazard to anyone wearing high heels and also be likely to cause unsightly staining.

8.5. SUPPORT TRESTLES FOR OUTFALLS

Surface water outfalls frequently cross the beach and need supporting to allow them to have clear discharge above beach level (Figure 8.18).

Choice of timber structure

Timber provides an ideal material for the construction of supports to outfalls since it is readily adaptable and easily worked on site.

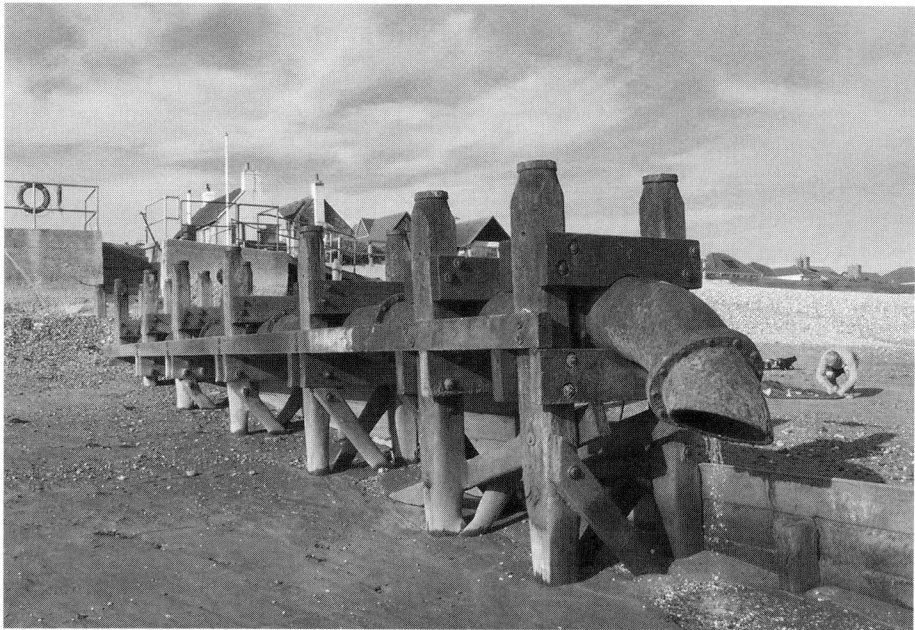

Figure 8.18.   Support trestle for outfall, East Wittering, West Sussex (courtesy Posford Haskoning)

Operating conditions

The operating conditions and selection of timber species for outfall supports are similar to those for groynes.

Overall configuration of structure

Outfalls are often located adjacent to groynes (Figure 8.19), enabling part of the groyne structure to be used to support the outfall and for the outfall to be sheltered by the groyne. Apart from the alignment, profile and length of the outfall itself, the configuration should also take into account the location of any joints or fittings along the outfall pipe.

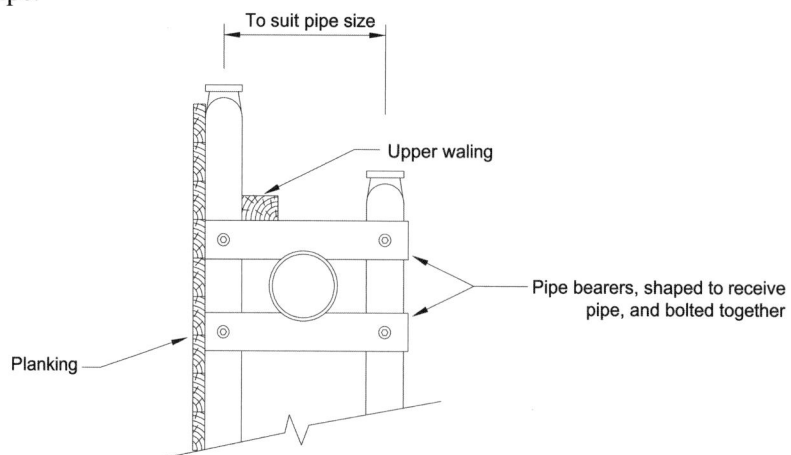

Figure 8.19.   Typical configuration for outfall supports adjacent to a groyne

Overall stability of structure

Overall stability is usually achieved by means of pairs of structurally linked piles, with diagonal bracing if possible.

8.6. ACCESS STEPS AND BOAT SLIPWAYS

Timber is often used to construct access from the promenade or riverbank onto the beach or into the water. Timber may present a slip hazard when wet, but is an attractive and durable material in many situations. See Figures 8.20 and 8.21.

Figure 8.20.   Ekki boat ramp at Tankerton, Kent (courtesy Mackley Construction)

Choice of timber structure

The main choice can be between an open timber structure and a solid concrete structure, although open composite structures of steel and timber or steel and concrete are not uncommon. Open structures have the advantage of lower impact on the natural movement of sediments and timber is often considered to be more aesthetically pleasing as well as resulting in less environmental damage.

Figure 8.21.   Purpleheart access steps at Eastbourne, East Sussex (courtesy HR Wallingford)

Operating conditions

Timber structures can usually work in situations where, at least for much of the time, boating and pedestrian activities are feasible. Allowance for seasonal or periodic variations in beach or bed levels should be made in designs; where conditions are particularly harsh, it may be possible to provide steps which can be removed during winter to avoid them being damaged.

Timber species

A significant consideration in selecting timber species for accesses is the likelihood and impacts of splintering. For this reason some tropical hardwoods (such as greenheart) should be avoided for treads, decks and handrails.

Overall configuration of structure

The main purpose of a slipway or a set of pedestrian steps is to facilitate access across a difficult terrain. This might include a significant change in level, an irregular surface such as a rock revetment or an unstable surface such as a shingle beach. In designing such structures it is necessary to identify what constitutes a difficult terrain and how it might change under natural conditions.

Opportunities are often taken to incorporate accesses adjacent to other structures, notably groynes on beaches (Figure 8.22). This can afford the access a degree of shelter and reduce abrasion as well as economising on the volume of timber required.

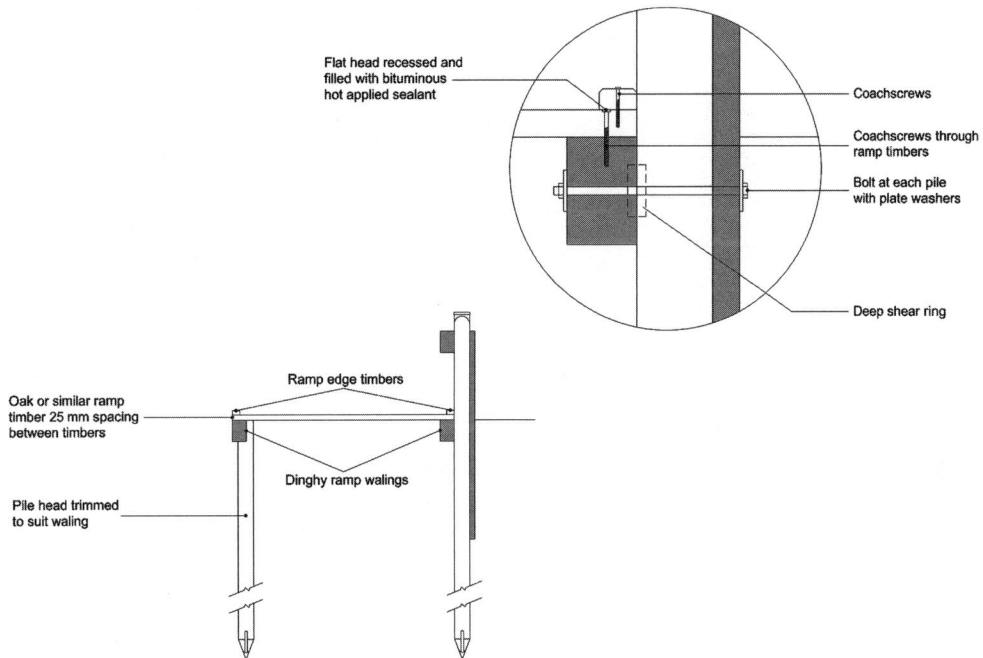

Figure 8.22. Dinghy (only) launching ramp at Shoreham Harbour built adjacent to a groyne

8.7. MOORING DOLPHINS, FENDER PILES AND RUBBING PIECES

Fender piles

These are usually deflection structures to prevent direct collision of ships with reinforced concrete or steel structures. Fender piles may be constructed from one or several piles, driven into the bed a short distance in front of the structure to be protected. The head of the pile is connected to the structure and a rubber buffer or spring is fixed between both. Another way of obtaining this spring effect is to drive single piles slightly tilted away from the structure and then pull the head towards the structure and anchor it.

Timber rubbing pieces

A range of timber species have been used successfully as rubbing pieces along jetties, piers, mooring dolphins and quay walls (Figure 8.23). The main requirements for rubbing pieces are the resistance to abrasion and splitting on impact. Rubbing pieces may easily be built-up of several timber sections, as the sections will mainly be loaded perpendicular to the joints. Elm, Douglas fir, pitch pine, oak, jarrah, opepe, ekki, okan are often-used species, frequently with a more durable timber mounted on a less durable species, this approach provides an economic solution whilst maintaining a sufficient thickness to prevent impact damage in the main structure.

Navigation structures, to guide vessels and avoid damage arising from impact between vessels and waterside structures, are constructed frequently from timber.

Figure 8.23.   Timber fenders at Axmouth Harbour (courtesy Posford Haskoning)

Dolphins

These are a cluster of piles tied together with horizontal and diagonal struts and bracings. They are used for moorings, as part of a lead-in jetty or to deflect shipping from navigation hazards such as bridge piers (Figure 8.24).

STRUCTURE SPECIFIC GUIDANCE

Figure 8.24.   Timber piled lead-in structure to protect bridge pier at Reedham, Norfolk (courtesy Clive Orbell-Durrant)

Operating conditions

Mooring piles and dolphins are vulnerable to accidental overloading. Care needs to be taken in selecting an appropriate operational load. Also, if possible, an appropriate failure mechanism should be designed into the structure.

Members and connections

Timber dowels, as opposed to steel fixings, can be an appropriate way of securing fenders and rubbing strakes to the parent timber structure. Dowels are more likely to expand, contract and deflect in harmony with the connecting timbers, resulting in a joint that remains tighter for a longer period of time. Also, any direct contact between the vessel and a timber dowel is less likely to cause a problem than a steel fixing.

Finishes and fittings

The actual navigational top mark, lantern, etc. is often a fitting to the main timber structure. The arrangement and detailing of such fittings should facilitate inspection, maintenance and replacement.

8.8. LOCK GATES, SLUICES AND FLAP GATES

8.8.1. Lock gates

Timber lock gates have traditionally been used on inland waterways (Figure 8.25) and still are used widespread. British Waterways have 1591 locks across their system, of which the majority are fitted with timber gates. The structure of timber lock gates is based on centuries of experience and therefore is difficult to improve.

Large lock gates may be constructed of steel, but often timber is used for the head and heel posts as it can be more easily fashioned on site to provide a water tight seal.

Figure 8.25. Timber lock gates on an upper reach of the River Thames (courtesy HR Wallingford)

Choice of timber structure

Timber has several advantages over other materials: it has a high strength/weight ratio, ability to withstand shock loads and ease of on-site repair. The majority of the locks on inland waterways in the UK are hand-operated. On these gates, balance beams are required for opening and closing the gate. Traditionally, these are large oak sections, e.g. 8.0 m long, but joining sections is possible. In some cases the mechanical control is fixed to the balance beam and can be disconnected for manual operation. The timber sizes required for the posts and planking are based on rather more experience than calculation, as loads on the gates are difficult to predict. Accidental loading may well exceed a hundred times the hydrostatic load on a gate.

Operating conditions

The life expectancy of timber lock gates is highly dependent on the operating conditions; particularly, the amount of traffic on the waterway and mechanical damage are influential factors. Generally, a lifetime of 30 years can be expected for hardwood lock gates. In some cases hard- and softwood are combined within one gate, constructing the frames of hardwood and the sheeters of softwood. The sheeters are replaced every 10 to 15 years and the frame may last for over 50 years. When replacing softwood sheeters for hardwood sheeters the balance beam should be fitted with weights to compensate for the decreased drift of the gate. Joints and fittings affect the lifetime of a gate, as decay of the timber is likely to take place in these areas first, and progressive failure will occur.

Timber species

Ekki, greenheart, some hardwood frames and softwood planking.

Overall configuration of structure (Figure 8.26)

The height of a gate is preferably larger than the width to avoid deformation.

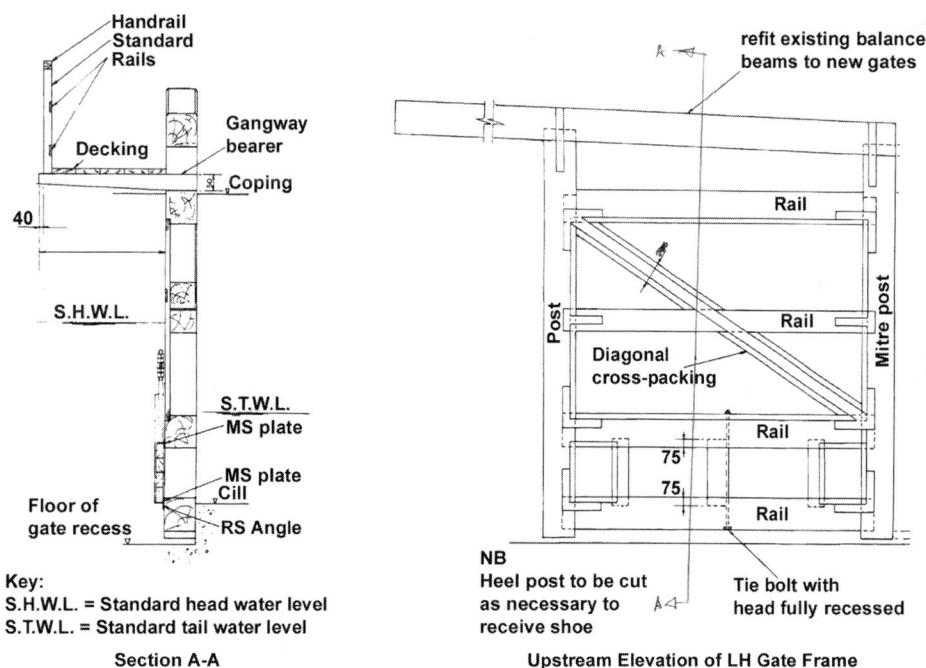

Figure 8.26. Typical conventional timber lock gate

Posts

Head/mitre posts usually have a larger width than thickness and are cut at an angle, from about 0.3 to 0.5 times their thickness, where the set of gates contact.

The width of the heel post may be larger than the thickness and is usually cut half round.

Members

The top and bottom members have larger section sizes than the ones in-between. The top member and members regularly above the water surface should be shaped to allow water to run off. The thickness of the intermediate members together with the sheeting is the thickness of posts. In some cases, a strut member will be placed in a line diagonal from the bottom of the heel post to the top corner of the mitre post to prevent the gate from distorting. For the same purpose a steel pull bar or strip can be fitted in the opposite diagonal. A bar will go through all the members and can be adjusted/tightened using a swivel placed above water at the top end of the heel post.

Sheeters

Sheeters can be oriented in several directions, the most common being in the vertical or diagonal direction. Placing the sheeters parallel to the strut member, if used, will support its function. A large sheeter width should be chosen in order to reduce the number of gaps.

Waling/fendering

To protect the gates from mechanical damage, timber fendering is attached to the waling.

Members and connections

Traditionally, the connections of members and posts of a gate are well-fitted mortise and tenon timber joints made by skilled carpenters. Steel plates bolted on either side of the timber sections support the joints. As the traditional timber joints are labour intensive to fabricate other possibilities have been sought. An example of a newly developed joint by British Waterways is presented in Figure 8.27. This steel joint requires less cutting of the timber. As the first joint of this design has only been fitted one year prior to writing, little can be said about its long term performance.

Curved washer plate

H section in 12 mm
plate rebated into timber

50 dia. SS dowel

20 dia. SS threaded rod

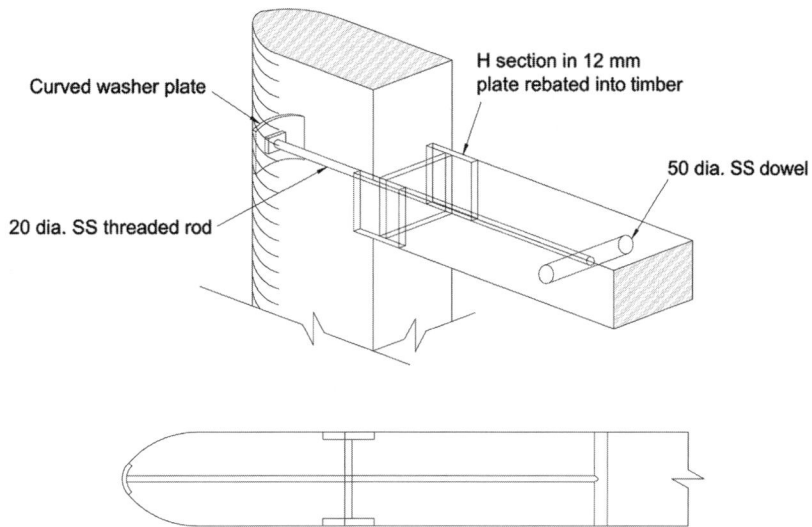

Improved Joint Detail

Figure 8.27.   New joint for timber lock gates developed by British Waterways (a) joint detail and (b) as implemented in a lock gate in the workshop (courtesy British Waterways)

The connection between gate heel posts and hollow quoins is an important detail, as this is where the gate hinges. When shut the connection should fit water tight, preferably causing as little abrasion as possible.

Lock gates are often fitted with walkways (see Figure 8.25) to allow pedestrians to cross the lock. Nowadays, gates tend to be fitted with aluminium prefabricated walkways, although in some cases the more aesthetic and traditional timber walkways are constructed on the gate. It should be borne in mind that a heavy walkway structure would increase the downward force and impart an extra torque on the gate. In some cases the balance beam may have to be weighted.

8.8.2. Sluices and flap gates

Figure 8.28 shows the mitre flap gates at the head of Lymington estuary which prevent saltwater flowing upstream into the Lymington river. These vertically-hinged twin-leaf gates (Figure 8.29) open when the fluvial head exceeds the tide level to allow river water to discharge into the estuary, and close when the incoming tide level rises above the river level. The gates are constrained from opening fully so that the water pressure from the incoming tide will push them closed.

The timber frames are constructed of ekki and the diagonal planking of pitch pine. Stainless steel brackets are used to reinforce the joints between the main structural members and vertical stainless steel rods are used to tie the main frame members at quarter-span points.

Hardwood timber lock gates are best left unpainted; this allows the moisture in the section to escape. Softwood finishing components such as walkways are best painted or impregnated with preservative because of their low resistance to decay.

Figure 8.28.   Lymington tidal flap gate (courtesy Babtie Group)

Figure 8.29.   Vertically-hinged, twin-leaf gate (courtesy Babtie Group)

Choice of timber structure

The operating environment for flap gates is one in which the gates will be constantly wetted and dried. The gates need to be strong enough to withstand the differential water heads that will occur and to carry the self weight of the gate from the hinge supports without undue distortion. The gates can be subject to damage, especially from floating debris in time of flood.

Timber can be relatively easily prepared to the required design and has the added advantage that both the closing posts and hinge posts can be finely sanded and smoothed in situ to provide good watertight seals. Planking can be replaced when it has outlasted its useful life.

Operating conditions

The conditions under which flap gates are required to operate, i.e. wet environment and repetitive hydrostatic loadings, are not conducive to a long design life. With the gradual deterioration and movement of the structure over time it is important that a method of adjusting the position of the hinge posts is incorporated in the design so that the gates are correctly set up and sealing faces are kept tight. To carry out repair work in the dry, a means of sealing off and emptying the gate channel is necessary. This may be achieved by the use of stop-logs, in which case stop-log grooves will need to be built into the channel walls upstream and downstream.

Timber species

The best choice of timber for the gate frame is a strong, durable hardwood such as greenheart or ekki. A softwood, such as pitch pine, is suitable for the planking.

Overall configuration of structure

The height of the gate frame will generally exceed the width. The main frame members are the hinge post, the closing post and the principal transoms. In the Lymington gates the main members are the full depth of the gates, 340 mm. The width of the hinge post is 340 mm, the closing post 280 mm tapering to 250 mm at the bottom, and the principal transoms 300 mm. The gates have three intermediate transoms which are 290 mm deep and 225 mm wide (upper two) and 300 mm wide (bottom). The planking is 50 mm deep, oversails the intermediate transoms, and is rebated into the main frame members. The diagonal planking runs from top outer corner to bottom hinge to help prevent sag of the gate.

The hinge post is supported on a cast iron pintle cast into the concrete floor slab. The pintle bears onto a socket recessed into the bottom of the hinge post. The top of the hinge post has a circular bronze band attached to it at top of wall height. A bracket rigidly fixed to the top of the wall has a strap attached to it which is wrapped around the bronze band. The strap can be tensioned to pull the top of the post closer to the wall to ensure that the seal is kept as watertight as possible.

The gates are set-up slightly off the vertical so that their selfweight, when not influenced by water pressures, acts to keep the gates in the closed position.

Overall stability of structure

The gates rely upon the fixity provided by the pintle at the bottom of the hinge post and the strap adjustment at the top of the post for their stability. It is important therefore to ensure that the pintle and strap adjustment bracket are properly connected to the channel structure and the forces imposed by the gates under load are taken into consideration in the design of the channel walls.

Members and connections

The main frame members are connected by mortise and tenon joints. The tenon connections at the end of the transoms are highly stressed and the timber stresses should be checked under maximum hydraulic loading conditions to ensure the tenon sizes are sufficient. All joints are reinforced by stainless steel plates on each face of the frame, which are bolted together through the structure.

Two vertical stainless steel rods are used on each leaf at quarter-span points to tie together the top and bottom transoms.

Finishes and fittings

All timbers have a machined finish. The hardwood timber frame is left in its natural state whilst the softwood planks have received a suitable preservative treatment.

Fittings are made from stainless steel except for some of the metalwork associated with the hinge post, which is made from cast iron or bronze as described above.

Stop logs

Stop logs may be used to close off locks during maintenance or to prevent boats from approaching a weir. Often long softwood logs are used, as is for temporary purposes, H-section steel wrapped in heavy-duty polyethylene.

Maintenance

Regular maintenance inspections should be carried out to ensure that the gates are sealing adequately, that they are closing under the correct differential heads, they are not obstructed by debris, and that they have not been damaged.

Adjustments can be made to the position of the top of the hinge post to improve the sealing with the channel wall.

To remove debris from between the gates and the cill, and to fully inspect the gates in the dry, a winching arrangement is required to pull the gates into their 'open' position. This can be achieved by connecting a wire rope from a winch located on the channel wall to a metal 'eye' bolted to the top transom.

Full inspection and repair of the gates can only be carried out by temporarily dewatering the channel and it is essential therefore that a method of stanking-off the channel is incorporated in the design. The normal method of achieving this is by the use of stop-logs, and stop-log grooves should be built into the channel walls for the purpose.

8.9. BED AND BANK PROTECTION AND STABILISATION

The structures discussed in this section use different material to those covered by the rest of the Manual. Whereas elsewhere the material of choice will be heartwood, here we are dealing with the use of sapwood (see Section 3.2 for explanation of the differences). In some cases, sapwood will comprise 100% of the material, in others it constitutes such a high percentage that the useful life of the structure is governed by its presence. The structures covered by it are:

Fascines (including faggots and wipes)	Submerged screens
Cribwork	Brushwood
Live timber stabilisation	Hurdles
Gabions	Stakes (driven or planted)

Choice of timber structure

There are a number of reasons for opting for timber structures which use sapwood.

- It may be available as a waste product arising from bank, farm or forest maintenance.
- It is an excellent habitat provider for a wide range of flora and fauna.
- It is aesthetically pleasing.
- Its production in quantity encourages traditional industries.
- It may be the only workable engineering solution.
- It may be required for heritage reasons.

Operating conditions

Almost invariably in Hazard class 5 (see Table 4.7).

Timber species

Willows and osiers *(salix* spp.*)* and alders *(alnus* spp.*)* should be used where rooting is required; otherwise any of the traditional coppiced timbers may be selected. Brushwood can be taken from any deciduous species.

Durability

Essentially these structures are not durable, however under certain conditions this can be improved. This is achieved by:

- Controlled harvesting. This should be done to optimise sap retention. As sap protects the growing wood from both insect and rot attack, its retention enhances durability (Box 8.3).

Box 8.3. Importance of cutting sapwood in due season

Both the durability and ease with which brushwood will strike can be greatly enhanced by skilful harvesting. The ideal time to carry this out is when the sap is rising in the wood to be harvested. This varies with the species and the weather, but is generally mid to late winter, i.e. well before the birds' nesting season. Some examples of this are:

1. A farmer had a large willow tree blow down one winter. He needed to replace a fence beside a wet ditch to keep cattle from falling in. He used the timber from the fallen willow to construct a post and rail fence. The posts took root, and he was left with a line of nicely spaced willow trees that lasted long after the rails had rotted away.

2. The Forestry Commission put up a rack of birch brooms near some isolated cottages so that the people living there had the means to hand to tackle wild fires before they became serious. After 18 months the broom heads were brittle and crumbling, and no use for their purpose. One of the cottagers cut some more birch twigs when their sap was rising, and re-headed the brooms. The new heads came into leaf (albeit briefly) for the next 6 springs, and they stayed supple for about 10 years altogether.

- Placing them in an anaerobic or low oxygen environment (Box 8.4).

Box 8.4. Durability in anoxic conditions

The Romans built a causeway across a bog in Yorkshire on fascines to carry a road. This stayed in use until the 1950s, when it started to show signs of distress due to the weight of modern traffic – well above the original engineers' brief! After unsatisfactory trials of a number of modern solutions, the local authority opened up some osier beds to produce new fascines to replace and reinforce the original construction.

Overall configuration and use of structures

1. Fascines, faggots and wipes. The word fascine is derived from the Latin *faces,* the bundle of rods surrounding the axe carried by the *Lictors* who preceded Roman magistrates and symbolised their powers of punishment; Roman engineers developed the idea as an engineering tool. The basic form of all these structures is the tied bundle. Modern fascines range in diameter from a few centimetres (willow wipes as stiffening and stone retainers to geotextiles in sea/river bed stabilisation) to several metres ('instant' crossings to anti-tank ditches). Bundles can be tied side to side to form mattresses. As an engineering tool they are underused, partly because many of the coppices and osier beds which produce the wood have fallen into disuse, so long lead times are needed to revive them before a major scheme can start. This does not conform well with current funding practices. They are also labour intensive to produce, which makes them expensive by comparison with, say, geotextiles in many applications. On the plus side, fascines are nearly carbon neutral, not subject to UV attack, robust, provide flotation and have a good weight to strength ratio.

2. Cribwork. This is the baby brother of a grillage. Criss-crossed poles are filled with stone to provide repairs to beds or banks. Brushwood should be included if the stone size is smaller than the poles to prevent them from washing out. On the riverbed the poles will need securing or weighting to keep them from floating. These are particularly useful for repairs to large washouts. Long term stabilisation is achieved by either sedimentation and/or plant roots binding the stone.

3. Live stabilisation. This is achieved by planting riverbank tree species to provide root stabilisation to banks. To provide initial repairs, suitable species (willow, alder) are woven hurdle fashion around stakes planted into the bed; these then root and grow to form trees. The best results are achieved when using wood cut with rising sap (Figure 8.30).

Figure 8.30.   Construction of a fascine mattress (courtesy Henk Jan Verhagen)

4. Gabions. The most traditional form of gabion is formed by planting uprights close spaced into the ground, then weaving thin wood between them to form a series of adjacent cells which are then filled with earth or stone; horizontal layers are often incorporated to give lateral stability. They were used to provide a stable core to sea and riverbanks from Roman times through to the nineteenth century. Military engineers also used them to provide protection against bombardment, usually in circular form. They are very labour intensive, and have been almost entirely replaced by wire mesh boxes in modern applications, but they may be a heritage requirement when ancient banks need repair.

5. Submerged screens. These are semi-permeable structures designed to deflect currents and encourage sedimentation. Typically, they consist of two rows of stakes planted or driven into the riverbed in-filled with faggots (small gabions) or brushwood. As it is usually undesirable to have them replaced by a line of trees, species such as willow and alder should be avoided for the stakes, as should wood cut with rising sap (see Box 8.3).

Figure 8.31. Accreting faggot, Holwerd, The Netherlands (courtesy HR Wallingford)

6. Brushwood. Apart from its use in conjunction with other structures, brushwood can be useful in its own right. Bundles ballasted and placed in riverbed butts upstream will form a snag and lead to the development of sand or mud bars. This application can also be used to perform minor bed repairs. Fixed to a fence in a stream they can be used to encourage sedimentation near banks to provide long term stability, and used with their butts driven into the bed to effect minor bank repairs.

7. Hurdles. Coppiced wood can be woven into panels known as hurdles. These can be fixed at any angle and act in a similar way to brushwood, but are more robust and can in addition form an effective 'rustic' fencing. They are normally fixed between posts and, assembled off site, can be quickly put up in an emergency and are easily handled where access is restricted. They can also be reused. Laid horizontally they form a traditional means of reinforcing earth embankments.

8. Stakes. Stakes, or round posts, are the usual method of providing strength and fixity to items 4-7 above. As they are largely sapwood, their durability depends on the same factors, and their useful life tends to be of the same order. They can also be used on their own, driven into the river bed to form lightweight moorings, as a screen to stop boats from approaching sections of bank, as supports for walkways, causeways and light jetties, and contiguously to form a temporary bank support or protection. In the latter case they are known as modular piles (Figure 8.32).

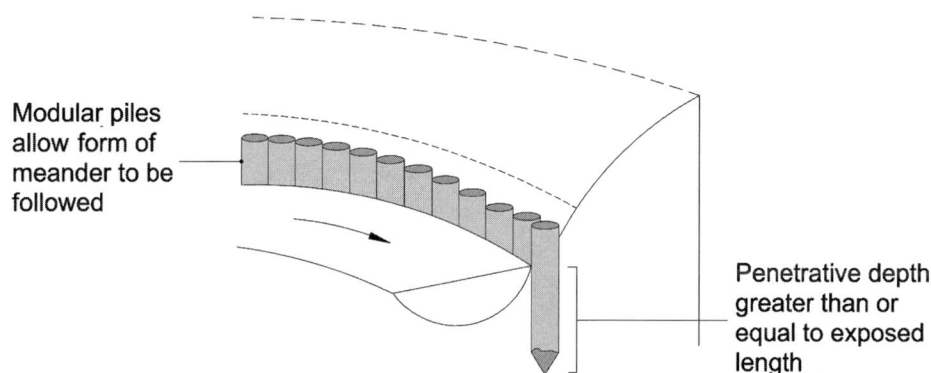

Figure 8.32. Modular pile bank protection (after Environment Agency, 1999)

Maintenance

By their very nature, these are very high maintenance structures. Growing structures can be controlled by grazing, browsing or pollarding, but otherwise the owner is faced with frequent inspection to remove debris which could cause immediate failure, especially in the case of the lighter structures, and may be faced with a total replacement situation as often as every two years if the bed or bank fails to stabilise. Careful and timely harvesting of the wood incorporated into the structures can reduce this commitment dramatically but, except in the case of structures in anaerobic conditions, maintenance will be much higher than it would be for tropical hardwoods in the same hazard conditions.

Literature

Environment Agency (1999). *Waterway bank protection: a guide to erosion assessment and management.* Environment Agency, Bristol, UK.

Graham, J.S. (1983). Design of pressure-treated wood bulkheads. *Proc. Coastal Structures '83*, pp 286-294, American Society of Civil Engineers.

8.10. EARTH RETAINING STRUCTURES AND CLADDING

This section includes:

- Toe protection
- Cladding
- River walls
- Composite cladding/walls

These form a continuum of cantilevered diaphragm structures, starting with simple cantilevered plank piles, and culminating in timber diaphragm held in position with anchored steel/concrete piles and walings. Cladding is normally trying to imitate a fully timber structure in appearance, and may contribute to the utility of the structure as well.

Choice of timber

These structures tend to be used where the consequences of failure would not be too serious, for instance:

- where there are no buildings or infrastructure behind the defence
- where the retained height is relatively modest
- where there is unlikely to be a high ground loading imposed behind the structure
- it is a cheap alternative where a wall needs to be built forwards a short distance
- to camouflage structures made of other materials
- to mediate between other materials and river users, e.g. timber facine to an 'H' pile can both act as a fender and protect the pile protection system.

Operating conditions

Although not without exception, these structures tend to be used in less aggressive conditions. They are more robust than those described in Section 8.9, but more vulnerable to failure than, say, a solid concrete or steel sheet-piled wall. Timber wall defences can be used in conjunction with soft defences to take the first impact of waves and currents and prevent undermining or uprooting of reed beds (rands) or grass slopes, especially in time of flood, or where there is a lot of seasonal boat traffic. Where ship wash is a serious problem, timber protection is unlikely to be adequate. Hazard class is usually Class 5.

Timber species

The selection of the most appropriate timber species relates to the following issues:

- strength of timbers, especially piles during driving
- stability of timbers with regards to splitting and distortion
- durability of members against decomposition, abrasion and marine borers.

Tropical hardwoods perform best in respect of the above criteria, but other species and/or treated timber should be considered, particularly where durability is not a prime consideration, the risks incumbent on failure are low or where they look out of place.

Overall configuration of structure

The posts and piles act in cantilever. A diaphragm is formed either with boarding fixed between the uprights, or driven in the form of plank piles (Figures 8.33 and 8.34). Where appropriate a waling can add extra support, and may also double-up as a rubbing strip or fender. The choice between the various options will depend in the strength of the hydraulic attack, the retained height and the use to which the particular length of riverbank is put. Where the main concern is undermining, a breastwork can provide toe protection. Toe erosion caused by wave or wake reflection can be a problem.

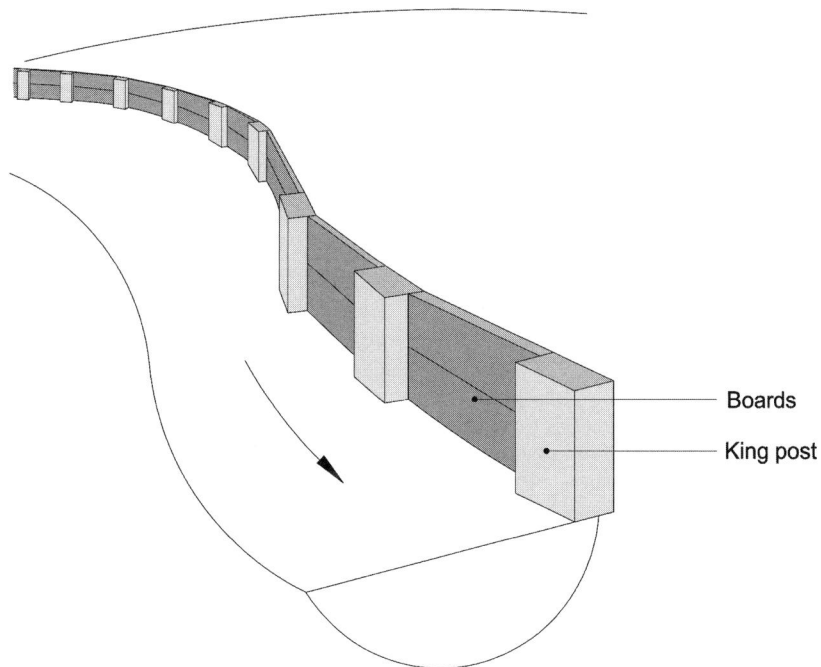

Figure 8.33.   Plank and post toe boarding (after Environment Agency, 1999)

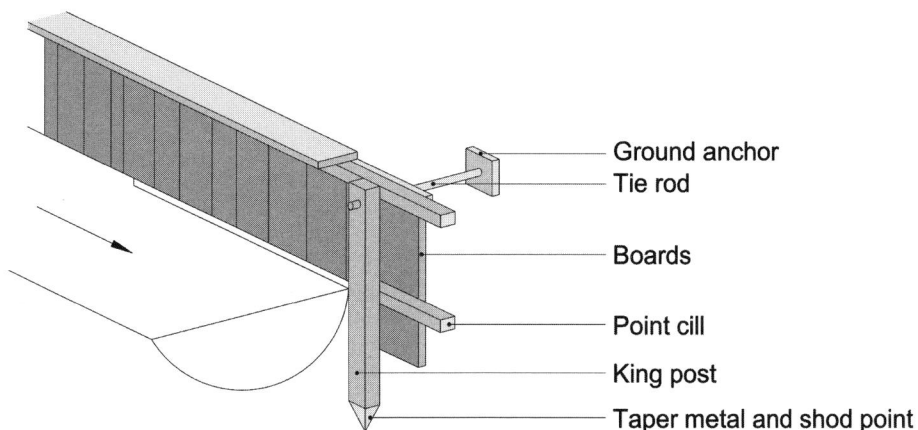

Figure 8.34.   Anchored timber piling (after Environment Agency, 1999)

Overall stability of the structure

Cantilever piles need to be driven to three to five times the retained height. Where substantial heights need to be supported, ground anchors or anchor piles or blocks with ties and walings may be required to provide sufficient support, especially in bad ground; additional support can be gained from buried planks (Figure 8.35). Anchored piles are usually shorter than full cantilever ones, as they normally need to be driven to only twice the retained height. All-timber construction is unlikely to be applicable where more than 3.5m of vertical support is required. When no anchors are used the ground near the wall will settle in order to achieve passive soil resistance below. When anchors are used these should be placed outside the potential failure circle. The use of a raking structure can improve the efficiency of a design.

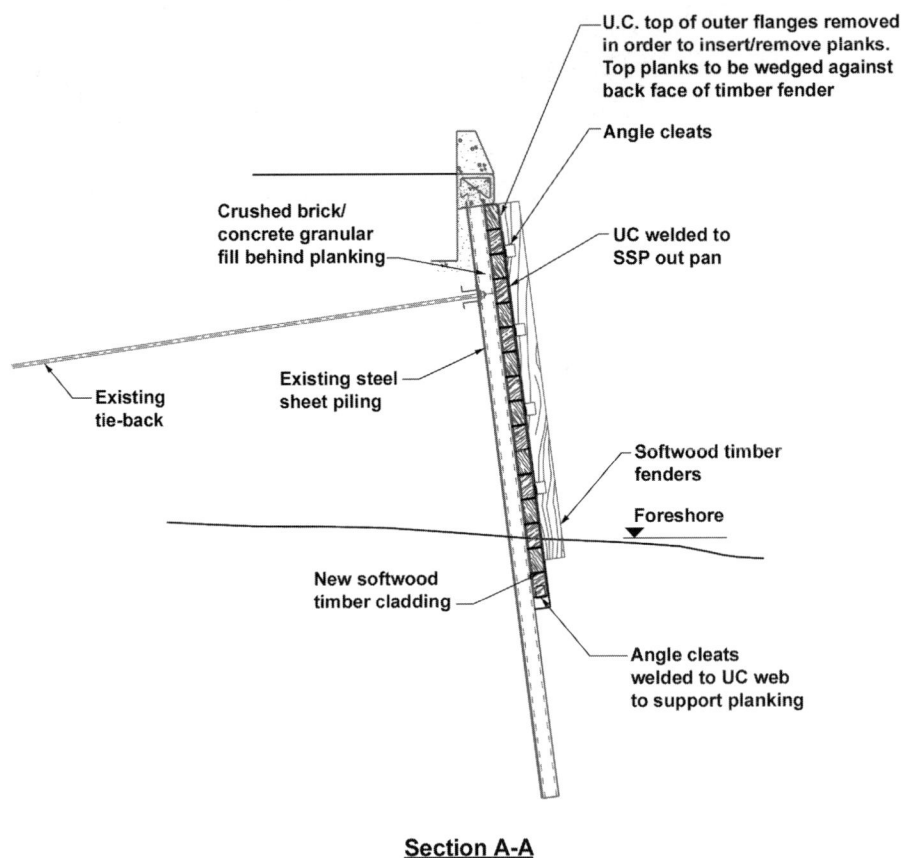

Section A-A

Figure 8.35. Typical section of river wall

Members and connections

These should be standard sections and lengths for ease of maintenance, generally to BS 5268. Fixings should be minimised, simple and designed to take account of any predicted impact loads and to last the anticipated life of the structure. Cladding fixings should be able to last the life of the underlying structure.

Box 8.5.   Case study, Deptford Creek (London)

At Deptford Creek, the Environment Agency returned to the use of timber as a primary material for the reconstruction of flood defences. One of the frontages of the creek consisted of a gradual slope backed by a derelict wall and an area designated to become an urban nature study area. It also contained buried cast iron services near the surface. An earth bund was proposed to replace the derelict wall, fronted by a regraded beach. Unfortunately, grading the front of the bund to the same slope as the beach would have placed the bund so far back that it would have put an unacceptable loading on the services, and severely curtailed the nature study area.

The solution was to stabilise the toe of the bank with a breastwork as shown in Figure 8.36 and steepen the front face, bringing the whole bank forward to improve both factors. The face of the bank and toe wall was then buried in coarse gravel scour protection.

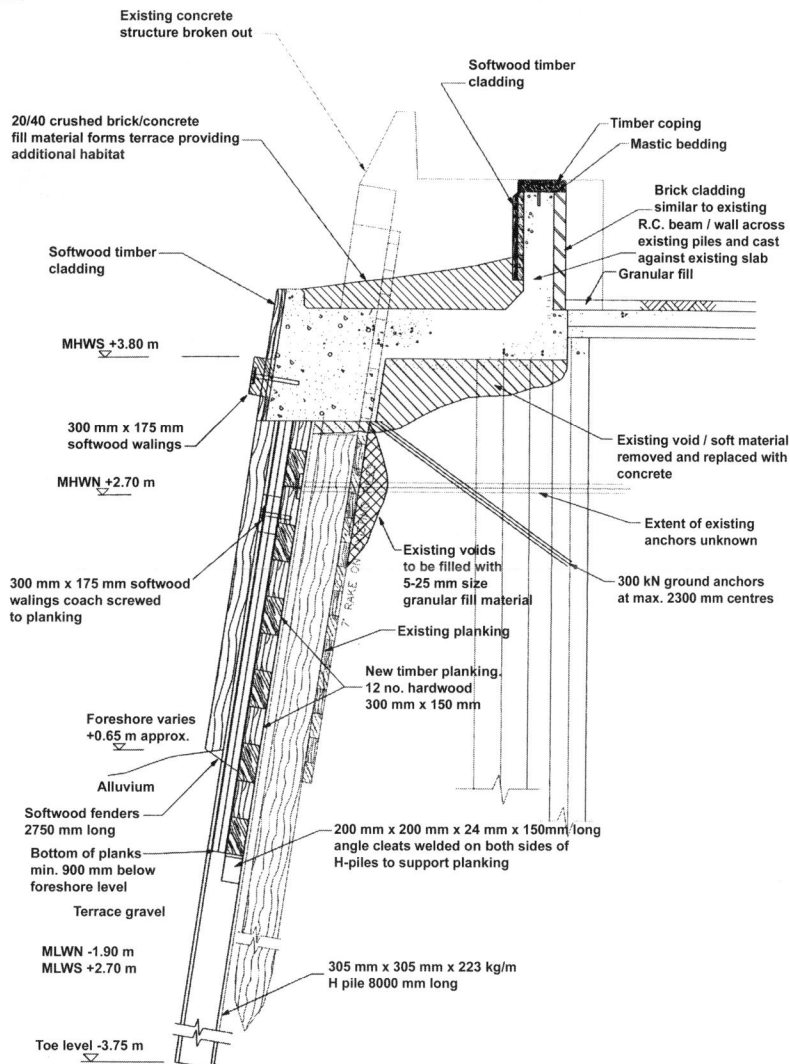

Figure 8.36.   King pile and plank wall at Deptford Creek

Box 8.5.   Case study, Deptford Creek (London) (continued)

The breastwork was constructed using steel king piles driven in front of the life expired timber wall, with ground anchors used to provide additional stability and hardwood timber planks spanning between the king piles. Timber had also been considered for the king piles, providing the main structural framework, but the required lengths and sections were not readily available and steel was used.

The selection of the structure and materials resulted from the consideration of the engineering, economic and environmental attributes of a wide range of alternative configurations. Opportunities for environmental enhancement were also identified and incorporated within the scheme. The works included breaking-out the existing concrete capping wall and setting back a new wall which was also undertaken to allow future raising if required.

The new king pile and plank wall is illustrated in Figures 8.36 and 8.37 which shows the main structural arrangement including the new terrace. Whilst the new wall does encroach marginally to the creek over part of the height, it also provides benefits resulting from the setback of the floodwall. The void between the new and old walls was filled with granular fill providing habitat for invertebrates. Hardwood was used for the structural planks but both the fenders attached to the front of the king piles and ledges or walings fixed at various heights to provide additional habitat opportunities were softwood.

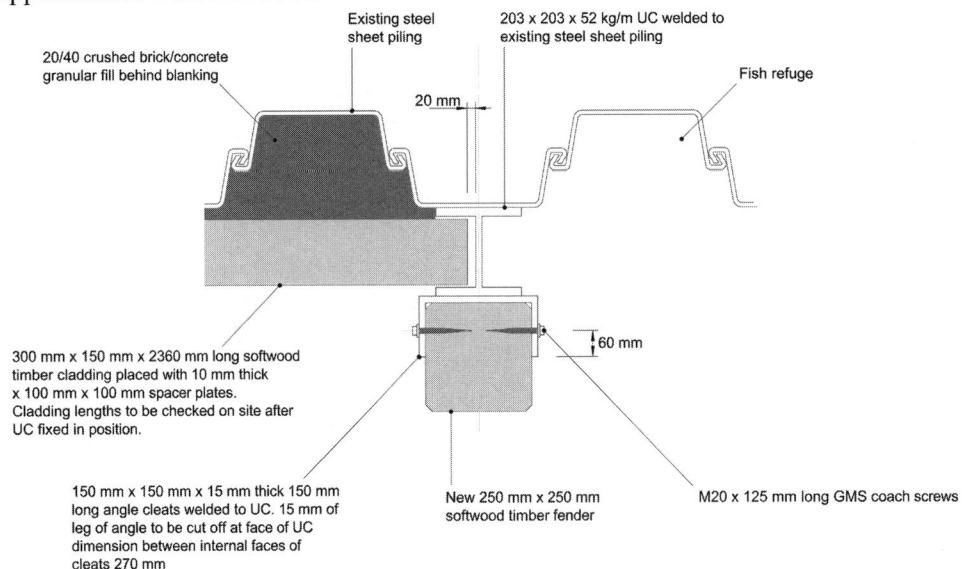

Figure 8.37.   Cross section and detail of cladding arrangement

The works also included timber cladding to existing steel sheet-pile walls, illustrated in Figures 8.38 and 8.39. Since the cladding is not structural, softwood was used and as with the king pile and plank wall opportunities were taken to provide a wide range of habitat enhancements including:

- bird roosts and nesting ledges (horizontal timbers attached to cladding at different levels)

Box 8.5.   Case study, Deptford Creek (London) (continued)

- opportunities for marginal vegetation (horizontal timbers attached to cladding at upper levels)
- fish refuges and spawning areas (openings in the cladding just above bed level to provide sheltered areas out of the main flow)
- invertebrate habitats (access to granular fill behind the cladding between cladding planks achieved by using spacer plates)

Figure 8.38.   Deptford Creek frontage before (left) and after (right) (courtesy Babtie Group)

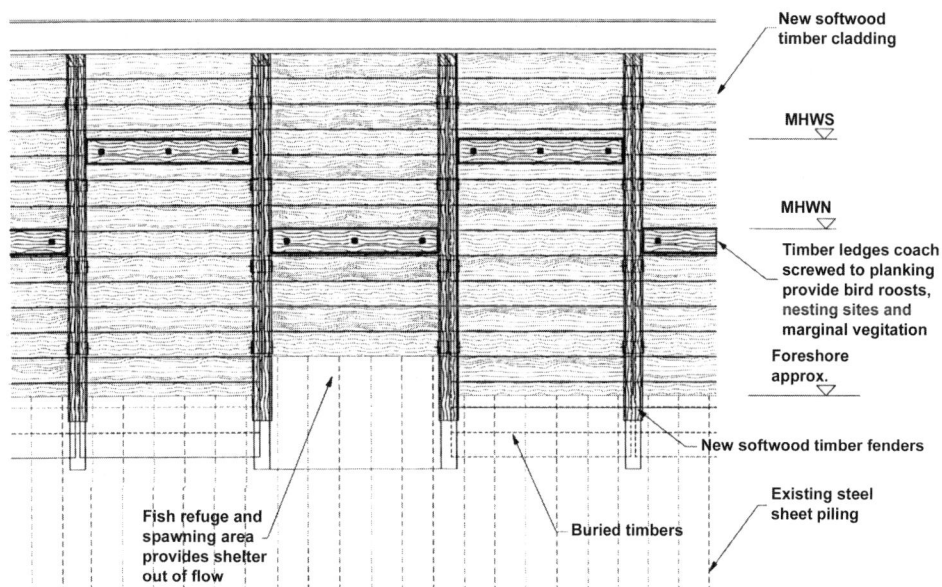

Figure 8.39.   Example of timber cladding from Deptford Creek

Finishes and fixings

The larger structures require bolts and coach screws; for smaller ones, timber spikes, screws or nails may be adequate. All should be either galvanised or stainless steel.

Maintenance

All timber structures need regular maintenance. This can be minimised by using the 'little but often' principle, whereby frequent inspection and reactive maintenance will identify and rectify defects before they get out of hand. Points requiring particular attention are physical damage, vegetation control, washouts behind the structure, loose fixings and decay of timber near the waterline. The cost penalty for poor maintenance is premature and expensive refurbishment. Cladding maintenance can be less onerous, as it can often be left until it becomes unsightly or a public hazard.

8.11. HABITAT PROVISION

One of the arguments for the use of timber is, that, as a natural material, it makes a better habitat provider than materials such as steel or concrete. Of course this can also be a drawback, as some of the beneficiaries, such as rot fungi, gribble, wood- and shipworm also accelerate degradation. Another beneficiary, which has a low popularity rating, is the brown rat (*Rattus norvegicus*). However, timber is easily adapted to meet site specific habitat requirements, such as:

- fish refuges
- bird perching, roosting and nesting sites
- plant pockets
- silt traps for invertebrates
- spawning shallows
- fry shelters
- tunnels for crayfish and water rats
- holts for otters
- and in decay can play host to a wide range of insect larvae, much to the benefit of their predators. See Box 8.5 for an example.

9. *Monitoring and assessment of timber structures*

The monitoring and assessment of timber structures is of critical importance to their effectiveness and durability. It is essential that any problem or deterioration is addressed quickly before it becomes significantly worse.

9.1. THE DEGRADATION PROCESS

Over the course of time and through the influence of the external environment, timber structures, like all other structures, experience a loss in performance, which is referred to under the umbrella title of degradation.

Degradation is both inevitable and undesirable. At best, degradation is limited to the surface layers of the timber structure. However, given the right conditions, degradation will eventually affect the whole structure, particularly in terms of mechanical performance, strength and durability. The rate of degradation is usually specific to the particular timber species and the environmental conditions of service.

The process of degradation can be subdivided into biological, chemical, photochemical, thermal, fire and mechanical degradation which are described in the following sections.

9.1.1. *Biological attack*

Fungal decay

Timber, provided it remains dry, i.e. below the decay threshold which is nominally 20% moisture content, will not suffer from fungal decay. However, timber in the marine and river environment will, in most cases, remain permanently wet. Where those timber elements are out of contact with water or the ground, they will probably be exposed to conditions that will result in them being vulnerable to intermittent attack (Figure 9.1).

Figure 9.1. Advanced fungal decay in primary and secondary beams (courtesy John Williams)

Fungal decay is both complex and variable and for it to occur, four conditions must be present: these are food, adequate moisture, suitable temperature and oxygen. In most circumstances, moisture is the limiting factor that restricts fungal activity. However, low levels of oxygen, limited nutrients and low temperatures also inhibit fungal activity. The timber itself or the cell contents provide the necessary nutrients. Oxygen is always available under service conditions except when the timber is totally waterlogged. Practically all fungal activity ceases at or below freezing point and is very slow just above it. Optimum conditions generally occur at around 20°C to 30°C and at 30% to 50% moisture content.

The fungal decay process may be loosely summarised as follows:

- Moisture content exceeds decay threshold.
- Bacterial, mould and sapstain attack. Mould and sapstain fungi are lower fungal organisms that are best described as 'scavengers'; they feed upon readily available cell contents but do not cause 'decay' of the timber structure. Their presence, however, indicates that conditions favourable for fungal decay exist.
- Soft rot attack occurs in ground contact and in marine and river environments, and tends to favour conditions where the timber is very wet. Soft rots are also 'lower' organisms.
- Dry rot attack is the most important destructive fungus found in buildings. However, with reference to timber structures in the marine and river environment, dry rot is extremely rare.

- Wet rots are by far the most important group of fungi to consider in terms of timber structures serving in the marine and river environment. Wet rot attack can be minimised/prevented by appropriate detailing as discussed in Section 7.3.

Insect attack

In general, timber structures serving in the marine and fresh water environment are not vulnerable to insect attack in the UK. However, there are exceptions to this statement and it should be qualified by the fact that sapwood regions that may exist in some timber elements will be attacked by wood-boring beetles. Ideally, untreated sapwood should not be present on these types of timber structure and where sapwood bands do exist, they will deteriorate rapidly. This may have consequences if fixings have been located within sapwood areas.

Where insect attack is noted over the timber structure, i.e. it is not limited to sapwood, it will usually be coincident with fungal decay. Such insect attack is commonly attributed to wharf-borer beetles and wood-boring weevil. Therefore, the presence of such organisms serves as a warning that more insidious organisms are at work degrading the structure.

Marine borers

All timber species are susceptible to attack by marine borers. The factors, which exert most control over the rate of attack, are density, silica content and extractives. Marine borers may also be restricted by environmental conditions such as temperature and salinity.

9.1.2. Mechanical degradation

Timber serving in the marine and fresh water environment will suffer mechanical degradation. By far the most common cause of mechanical deterioration of timber occurs when it is stressed under load for prolonged periods of time – this is known as creep. Creep may be defined as a loss in strength with time under load. For example, for a timber element to sustain a load for 100 years, then the load must not exceed 50% of the load the timber would fail at in the short term. In all structural applications, a load endurance factor should be incorporated in the design and is usually applied as one of several contributing parts to an overall safety factor.

In the marine and fresh water environment, tidal action, beach abrasion and impact can all affect the integrity of timber structures. With reference to impact damage, this can result in compression failure of the timber fibres. Abrasion can be a major factor in determining the life of a timber structure and may be most apparent in areas where there is a small tidal range as the erosion will be concentrated on a relatively small part of the structure.

9.1.3. Chemical degradation

The most important characteristics relating to chemical resistance are impermeability and density. Chemical attack is restricted to the surface of the timber and can be caused by acid or alkaline chemicals. Degraded wood surfaces have a 'fluffy' appearance, i.e. defibrillation of the wood fibres, and are quickly eroded. The attack is superficial. In the marine environment, sea spray results in salt deposit forming on the wood surface which can result in defibrillation of the fibres. One positive side effect of these salt deposits is that they act as a natural preservative and discourage many species of fungi from degrading the timber.

9.1.4. Photochemical degradation

On exposure to ultra-violet light, the coloration of the heartwood of most species (such as teak, mahogany and iroko) will lighten although a few species such as Douglas fir will actually darken. Colour change is part of the photochemical degradation process. The effects of wind and rain also cause complex degradation at the timber surface. Weathering results in the loss of surface integrity of the timber and affects its appearance. Generally, weathering does not the affect the long term performance and strength of timber elements. There is one exception and that is the breakdown of surface finishes, i.e. paint films, can result in premature failure of joinery products as water penetrates through the paint film and elevates the moisture content of the component to levels which can support fungal decay.

The rate of photochemical degradation and weathering results in slow erosion of the timber surface and can act as a precursor to fungal decay. With reference to maritime and fresh water structures comprising dense, large sections of timber, weathering is insignificant.

9.1.5. Thermal degradation

Very low temperatures have no effect on the strength properties of timber. High value softwoods originate from well within the Arctic Circle where for many months of the year the tree is standing in permafrost. Thermal degradation only occurs in timbers that are exposed to high levels of radiant heat where the surface can attain a temperature above 100°C. The surface of the timber will darken.

9.1.6. Degradation by fire

Timber is an organic material, and as such is combustible in the dry state (Figure 9.2). The thermal energy required to ignite and sustain the combustion of timber is considerable. Timber has the ability to form a char layer, which is a good insulator that retards combustion. The behaviour of timber in fire is predictable in relation to the section size of the component and its density.

Fire degradation in marine structures may be viewed as uncommon and can occur by accident or by vandalism. Where fire degradation has occurred, very often it is possible to predict the residual strength of the structure by considering the reduced sections of timbers which have been protected by the char layer.

Figure 9.2. Fire damage down to the water line of Southend Pier (courtesy Ecotimber Ltd.) – see also Section 10.2

9.2. MONITORING AND ASSESSMENT

Timber structures should be monitored and surveyed as part of a regular maintenance regime. The frequency and extent of the surveys will be dependent upon the level of risk the structure is exposed to, and also on previous survey results. For example, a timber structure situated in a location where there is a high incidence of marine borer would have to be monitored at more frequent intervals than a comparable structure located where there is a low risk of marine borer.

Furthermore, a structure made of a moderately durable timber would, generally, have to be monitored more frequently than a structure made out of a very durable timber.

There are a number of methods used by the surveyor to assess the condition of a timber structure and to enable the surveyor to estimate its potential life span. These include visual and other non-destructive (such as ultrasound examination) techniques as well as intrusive techniques such as drilling or probing.

9.2.1. Inspection of timber structures

There are five factors which must be considered when inspecting timber structures:

- the timber species used in the original construction and the species of any replacement components
- evidence of preservative treatments – this may be determined both quantitatively and qualitatively
- identification of the presence and severity of mechanical and/or biological degredation
- appraisal of the condition, size and location of the relevant timber components
- condition of fixings and joints. Poor detailing of joints and selection of inappropriate fixings can significantly shorten the life of the structure.

Having carried out a preliminary visual inspection of the structure, the fastest method of surveying any timber structure is to use hammer soundings to determine the resonance. The timber components are struck at regular intervals consistent with force applied in the hammer blow. This technique provides an indication of gross deterioration. Large pockets of decay are coincident generally with areas that sound hollow when struck. Fissuring, mortising and notching of the timber can influence hammer soundings, so the surveyor should be aware of these features that may influence the findings.

Decay detection drilling may be used to determine the condition of timber components. However, when surveying timber structures that have been constructed from heavy, dense timbers such as ekki, greenheart and balau, this method is extremely time-consuming. This is a particular disadvantage when working within short time windows governed by the prevailing tides. Specialist drilling equipment is often used. Such apparatus can produce a printout of the timber's resistance to a sharp needle-like probe applied to the face of the timber using a constant force. The printout illustrates a profile of the timber components' hardness. Significant changes in hardness, illustrated on the printout, may indicate the presence of fungal decay or significant marine borer attack. The surveyor must also be aware that fissures and the presence of pith may also affect the printout.

A faster method of decay detection drilling is to use wood augers of typically 6 mm diameter and an electric drill. Although this method does not produce a printout of the areas surveyed there are obvious advantages with this method when time on site is limited by tidal factors. This method also requires a high degree of surveyor competence and experience. It is important to carry out the drillings in a consistent manner, i.e. apply constant force when drilling and ensuring the wood augers are replaced frequently.

In conjunction with drilling, the surface of the timber should be probed with a sharp screwdriver or bradawl. This will enable the surveyor to identify areas of surface softening, and possibly, severe deterioration. It is particularly appropriate in areas around joints and fixings should be carefully surveyed by this method with minimum damage to the structure.

Increment core samples may be taken from the timber structure using a Pressler borer. This technique enables the surveyor to assess the condition of the timber under laboratory conditions. Microscopic analysis may reveal fungal activity. These core samples may also provide sufficient material for species identification and preservative analysis. It should be borne in mind that core samples represent a very small proportion of the timber structure. Taking core samples can prove to be a valuable exercise where the condition of the inaccessible surfaces of decking boards, for example, need to be determined. Specialist timber-coring equipment should be procured so that core samples may be extracted from dense tropical hardwoods.

To determine the residual strength and life span of any timber component it is important to know the original dimensions of the component, its current dimensions and its condition. This is often possible in cases where the timber structure has been exposed to attack by soft rot fungi and gribble, as the attack is superficial. Determining the rate of deterioration should enable the surveyor and engineers to estimate the remaining useful life expectancy by estimating the time it will take for the structure to erode down to its minimum serviceable cross-section area.

However, it is important to recognise the type of deterioration affecting the structure and to categorically define the full extent of fungal decay, if present. In some cases, this can prove to be difficult as the early onset of decay by wet rot fungi, for example, may not always be detected by survey methods. Therefore regular inspections of the structure should be incorporated as part of any maintenance regime. As fungal decay develops, it usually does so at an ever increasing rate with the result that the subsequent loss in strength of the component also occurs at an ever accelerating rate. Where attack by shipworm has occurred, estimating the residual strength and life span of affected components is notoriously difficult and would involve detailed, destructive assessment of a number of the timber components.

In general terms, the following points may be considered by the experienced timber surveyor.

- In all cases, where the timber element is below the riverbed or beach level, the timber will remain relatively free from biological attack. The greatest hazard is at the intertidal zone for the marine environment and between the high and low water marks for river environments.

- In cases where the timber is uncovered, out of contact with water and out of ground contact, the Hazard class HC3 will apply. The timber may be at risk of fungal decay intermittently and care should be exercised when surveying, especially around joints and horizontal shaded surfaces where moisture can collect.

- Timber in ground contact is particularly vulnerable at the interface between the soil and air. As the ground surrounding the timber element is excavated to greater depths, the likelihood of finding external fungal decay diminishes due to the lack of oxygen. For example, the ground around timber posts is usually excavated to a depth of 0.5 m and the pole surveyed for decay.

9.3. IDENTIFICATION OF TIMBER SPECIES

The importance of correctly identifying the timber species has already been stated. Although different species of timber may be assigned the same *visual* grade, they are not assigned to the same *strength* class. For example HS grade iroko may be assigned to strength class D40 whereas HS grade greenheart may be assigned to strength class D70.

At first sight, the identification of timber appears to be a relatively straightforward exercise. However, literally thousands of species exist and, very often, correct identification is an extremely difficult and complex task, best carried out by an experienced anatomist or timber technologist. In some cases, it is not possible to identify a specific species although it may be possible to narrow it down to a species group. Two examples of species groups are given below.

- Southern yellow pine is the commercial term for several closely related species including *Pinus palustris* (longleaf pine), *Pinus elliotii,* (slash pine) *Pinus echninata* (shortleaf pine) and *Pinus Taeda* (loblolly pine). It is not possible to differentiate between these timbers using anatomical features of the timber although differentiation is possible on botanical features for example, needle and cone sizes.

- Meranti, seraya and lauan are the commercial names given to a large number of species of the genus *Shorea* distributed throughout Southeast Asia. Generally, the name 'meranti' applies to timber from Malaya, Sarawark and Indonesia. Seraya applies to timber from Sabah, and Lauan for timber from the Philippines. The timbers may be further grouped by colour and density but it is often not possible to distinguish individual species.

Most end users of timber handle a narrow range of species and are able to identify the timber by briefly examining the gross features of the wood combined with an appreciation of the member's mass. Typical gross features are:

- colour
- mass
- odour of freshly cut surfaces
- texture and grain
- conspicuous rays and vessels.

This approach is suitable where interest is confined to a limited species range. However, where identification requires a higher degree of assurance, samples may need to be examined using a hand lens or a microscope and computer aided identification keys. This will normally require one or more samples (typically, each the size of a matchbox) to be provided for a specialist to examine in detail. When taking such samples care must be exercised to ensure that they are representative of the structure as a whole rather than being selected from parts of the structure that are in the best or worst condition.

10. Maintenance, repair and enhancement of timber structures

Timber structures, typically, require a relatively high commitment of maintenance and repair activities. It is important that this is budgeted for at the outset and considered within the design approach and selection of materials.

10.1. DESIGN PROCESS

In addition to ensuring that design details contribute to the durability of timber structures, it is important that provision is made for necessary maintenance and repair works. The design process should actively consider the likely works required over the whole duration of the scheme lifetime and ensure that maintenance activities can be undertaken safely and efficiently. This may include steps such as using stainless steel bolts and coach screws which are far more durable than mild or galvanised steel and, consequently, much easier to undo after a few years exposure to the sea. Similarly, where different elements of a structure have different service lives (such as softwood planking on a hardwood frame, or the planks and piles of a groyne) consideration must be given to the replacement of the less durable components. Where elements in a particular location suffer greatest degradation (such as the high waterline on lock gates), the design of other components (in this case the frame) may be altered to minimise the quantity of material replaced.

The design process should seek to attain:

- the lowest whole life costs, provided that the required functionality is retained
- the least impact on the environment in maintaining and operating the asset, including the sourcing of the construction materials used in any repairs
- retained functionality whilst maintenance and repair of specific assets is carried out
- predictable performance under the impact of an extreme event beyond the design standard
- ease of repair (readily accessible connections joints and minimum wastage when replacing elements)
- minimal negative impacts and the maximum benefits to society over the life of the asset.

Careful detailing during the design and construction phase can significantly reduce the overall maintenance commitment for some timber structures.

10.2. MAINTENANCE ACTIVITIES

Maintenance of timber structures may be undertaken for a number of reasons including:

- replacement of damaged or degraded elements (see Figure 10.1)
- adjustments to the structure profile or configuration to increase effectiveness (for example, adjusting the height of a timber groyne)
- to ensure public safety
- avoidance of further damage (for example, loose planking on a groyne or lock gate vibrating excessively and causing damage to primary and/or secondary structural elements).

Figure 10.1. New timber fixed onto old as part of refurbishment of fire damaged Southend Pier (courtesy Southend Borough Council) – see also Section 9.1.6

Most maintenance regimes combine routine inspections with reactive repairs. This avoids excessive costs and possibly unnecessary work which could result from strictly programmed maintenance activities, but still ensures that defects can be speedily identified and remedied. With experience it is often possible to predict where maintenance work will be needed and even sometimes to put proactive measures in place (see, for example, Box 10.1).

All timber structures will require the bolts and coach screws to be tightened from time to time in service. This is usually the result of shrinkage in service if the timber is above the intertidal zone. Other factors that can result in loosening of the fixings can be local yield at points of high stress or reversal of loading as temporary supports are removed.

Box 10.1. Use of plywood backing to groyne planks at Bournemouth

Bournemouth has a field of 50 impermeable timber groynes, typically 75 m long at 180 m spacing. The piles are 300 mm × 300 mm, and the planks are 300 mm × 100 mm, and predominantly greenheart. Six groynes were planked with ekki, one with balau, and one with jarrah in the mid 1980s. The beach is a fine sand, with an appreciable shingle content.

Older groynes display slight gaps between their planks. Sand transport through the gaps is considerable, and abrasion increases the width. Shingle becomes jammed in the opening, preventing closure when the timber expands in the summer months. Sand transport increases when the gaps reach about 10 mm, and beach levels fall against the groyne.

A simple, cost-effective form of repair is the addition of 'tingles'; marine ply cut to 2400 mm × 100 mm × 38 mm and nailed over the gaps with stainless steel ring-shank nails. Tingles have a useful life of two to five years but are cheap and easy to replace.

New groynes also display gaps of 1 to 3 mm between brand new planks because of dissimilar widths or bows. Whole sheets of marine ply are fitted to the sheltered side of the planking, to seal the gaps and prevent the onset of erosion. The first panels were fitted with ring shank nails, but a few panels came loose in storms, and bolts were used instead (see Figure 10.2). Later, coach screws were used to avoid drilling holes through the greenheart planks. The oldest panels have been in place for seven years, rendering the groynes completely impermeable. This delays the onset of inter-plank abrasion, and extends the life of the greenheart planks.

Gribble has been increasingly noted at Bournemouth in recent years, infesting the edges of a minority of greenheart planks in the intertidal zone. Some planks were severely damaged in just seven years. A few panels have been removed for inspection, to ensure that they do not increase gribble attack on the sheltered side of the greenheart, or the ply itself.

Two types of plastic panels have been used, to eliminate the risk of gribble attack.

Figure 10.2. Use of plywood backing to groyne planks at Bournemouth (courtesy Bournemouth Borough Council)

An example of the maintenance regime for timber groynes adopted by a local authority on the south coast of England is provided in Box 10.2. It is notable that focused, relatively low cost maintenance activities can have a significant impact on the longevity and durability of timber structures, yielding considerable economic benefits. One example of such works is the use of rubbing pieces on the piles of timber groynes to protect them from abrasion.

Box 10.2. Repairs and maintenance of timber groynes: Arun District Council

Arun District Council recognises abrasion as the main problem on their groynes. The two main areas of abrasion are on the piles, at beach level on the down-drift side caused by the wave run-up and back wash, and immediately above planking level where shingle is moved over the groyne by the wave action.

Beach levels are allowed to settle for approximately two years following installation of the groyne and then an assessment is made of the need and location for 'pile rubbing strips'. These are sacrificial lengths of old planks or offcuts of new planking coach screwed or bolted to the pile in a suitable location to bear the brunt of the worst abrasion. These can then be replaced as necessary at a later date or relocated to a new area on the pile to ensure there is no abrasion to the main pile structure.

General maintenance of groynes is facilitated by biannual inspections, which feed into an annual maintenance contract with the works normally carried out by the in-house works section. The inspections are supplemented by *ad-hoc* inspections

---

> **Box 10.2. Repairs and maintenance of timber groynes: Arun District Council** (continued)
>
> carried out during other works and by information from the public. We tell the local residents that they are our eyes on the beach and they realise that early replacement of a lost or broken plank is generally in their interests.
>
> Generally, maintenance consists of replacement of missing or broken planks, fixing of additional planks as the beach profile is improved and the fixing of pile rubbing strips. Gradual raising of plank levels also spreads any abrasion over a wider area of the pile hence reducing the overall impact.
>
> There are approximately 300 timber groynes on the Arun District Council frontage and the coast protection maintenance budget is in the region of £70 000 per annum. This budget is provided to cover all maintenance costs, including timber supply and execution of the works. Of that annual budget, approximately £40 000 will be utilised on timber purchase and timber groyne repairs and maintenance.

10.3. END OF LIFETIME

At a certain point a timber structure will have to be replaced or dismantled and reconstructed, usually because it presents a significant safety hazard or maintenance becomes increasingly uneconomic. The timing of this is usually determined by the condition of the primary members, which are the most difficult and expensive to replace.

Once a structure is deemed to have reached the end of its useful life it is important that a thorough assessment determines the most appropriate replacement, taking care to learn lessons from the performance and condition of the previous structure. Depending on the structure in question it may be beneficial to construct the replacement before dismantling the original, as this minimises disruption of the function provided. However, it could make direct reuse of substantial quantities of timber difficult, as this will need to be stored until the next work is undertaken.

The life expired structure should be carefully dismantled rather than demolished, with any opportunities to recycle materials carefully explored and minimal damage caused to the substrate. Waste timber can normally be recycled in some form or used for energy recovery, but great care must be taken with preservative-treated timber components. It should be noted that preservative-treated timber is now classified as hazardous waste and therefore:

- cannot be burnt
- carries a premium when disposed of to tip.

These factors need to be taken into account when carrying out whole life costings against untreated hardwoods. There is some research into the possibility of recycling, but results will be too late for this publication.

10.4. SELECTION OF TIMBER

The selection of appropriate materials for the repair and adaptation of timber structures is largely determined by the original materials and anticipated residual life of the remaining structural components. It is often prudent to keep a stock of appropriately dimensioned timbers to allow groyne or deck planks to be repaired with the minimum of delay.

11. References

Baharuddin, H.G. and Simula, M. (1996). *Timber Certification in Transition.* International Tropical Timber Organisation, Kuala Lumpur.

Brookes, A. and Agate, E. (2000). *Sand dunes: a practical handbook.* British Trust for Conservation Volunteers http://handbooks.btcv.org.uk/handbooks/index/book/137

BS 144: 1997 *Specification for coal tar creosote for wood preservation* (AMD 9947).

BS 373: 1957 *Methods of testing small clear specimens of timber.*

BS 648: 1964 *Schedule of weights of building materials* (AMD 105) (AMD 344).

BS 1282: 1999 *Wood preservatives – Guidance on choice, use and application.*

BS 4072: 1999 *Copper/chromium/arsenic preparations for wood preservatives.*

BS 4169: 1988 *Specification for manufacture of glued-laminated timber structural members* (Partially superseded by BS EN 392).

BS 4978: 1996 *Visual strength grading of softwood* (AMD 9434).

BS 5268-2: 2002 *Structural use of timber. Code of practice for permissible stress design, materials and workmanship.*

BS 5268-5: 1989 *Structural use of timber. Code of practice for the preservative treatment of structural timber.*

BS 5450: 1999 *Round and sawn timber – permitted deviations and preferred sizes.* Hardwood sawn timber.

BS 5589: 1989 *Preservation of timber.*

BS 5666-1: 1987 *Methods of analysis of wood preservatives and treated timber. Guide to sampling and preparation of wood preservatives and treated timber for analysis.*

BS 5666-2: 1980 *Methods of analysis of wood preservatives and treated timber. Qualitative analysis.*

BS 5707: 1997 *Specification for preparation of wood preservatives in organic solvents.*

BS 5756: 1997 *Visual strength grading of hardwood.*

BS 5930: 1999 *Code of practice for site investigations.*

BS 6349-1: 2000 *Maritime structures. Code of practice for general criteria.*

BS 6349-6: 1989 *Code of practice for maritime structures. Design of inshore moorings and floating structures.*

BS 6399-1: 1996 *Loading for buildings. Code of practice for dead and imposed loads* (AMD 13669).

BS 6399-2: 1997 *Loading for buildings. Code of practice for wind loads* (AMD 13392) (AMD Corrigendum 14009).

BS 6399-3: 1988 *Loading for buildings. Code of practice for imposed roof loads* (AMD 6033) (AMD 9187) (AMD 9452).

BS 7359: 1991 *Nomenclature of commercial timbers including sources of supply.*

BS 8004: 1986 *Code of practice for foundations.*

BS 8417: 2003 *Preservation of timber – recommendations.*

BS EN 275: 1992 *Determination of the protective effectiveness against marine borers.*

BS EN 301: 1992 *Adhesives phenolic and aminoplastic for load-bearing timber structures. Classification and performance requirements.*

BS EN 335: 1992 *Hazard classes of wood and wood-based products against biological attack.*

BS EN 338: 1995 *Structural timber. Strength classes.*

BS EN 350: 1994 *Durability of wood and wood-based products. Natural durability of solid wood.* Part 1. Guide to the principle of testing and classification of the natural durability of wood. Part 2. Guide to the natural durability and treatability of selected wood species of importance in Europe.

BS EN 351-1: 1996 *Durability of wood and wood-based products – Preservative-treated solid wood.* Classification of preservative penetration and retention.

BS EN 386: 2001 *Glued laminated timber – Performance requirements and minimum production requirements.*

BS EN 408: 1995 *Timber structures. Structural timber and glued laminated timber. Determination of some physical and mechanical properties.*

BS EN 460: 1994 *Durability of wood and wood-based products - Natural durability of solid wood – Guide to the durability requirements for wood to be used in hazard classes.*

BS EN 519: 1995 *Structural timber. Grading requirements for machine strength graded timber and grading machines.*

BS EN 599: 1997 *Durability of wood and wood-based products – Performance of preventive wood preservatives as determined by biological tests.*

BS EN 844-1: 1995 *Round and sawn timber.* Terminology. General terms.

BS EN 1313: Part 1 (softwoods):1997 *Round and sawn timber – permitted deviations and preferred sizes.*

BS EN 1313: Part 2 (hardwoods): 1999 *Round and sawn timber – Permitted deviations and preferred sizes.*

BS IS 15686 *Service life planning of buildings and constructed assets.*

BS EN ISO 1461: 1999 *Hot dip galvanised coatings on fabricated iron and steel articles – specification and test methods.*

BS EN ISO 14001 *Environmental Management Systems. Specification with guidance for use.*

BWPDA (1999). *The British Wood Preserving and Damp-proofing Association manual.* The British Wood Preserving and Damp-proofing Association.

Certified Forest Products Council (2002). Available from www.certifiedwood.org

CDM (1994). *The construction (Design and Management) Regulations Statutory Instrument 1994 No. 3140.* London: the Stationery Office.

CIRIA (1996). Beach management manual. Simm, J.D., Brampton, A.H., Beech, N.W. and Brooke, J.S. (eds.). Construction Industry Research and Development Association, Report 153.

Council Directive 76/769/EEC of 27 July 1976 on the approximation of the laws, regulations and administrative provisions of the Member States relating to restrictions on the marketing and use of certain dangerous substances and preparations.

CP 3: Chapter V-2: 1972 *Code of basic data for the design of buildings. Loading. Wind loads* (AMD 4952) (AMD 5152) (AMD 5343) (AMD 6028) (AMD 7908) (No longer current but cited in building regulations).

DD ENV 1995-1-1: 1994 Eurocode 5. *Design of timber structures. General rules and rules for buildings.*

Defra (2001). *Government timber buying power to help combat illegal logging.* Department for Environment, Fisheries and Rural Affairs. Available from http://www.defra.gov.uk/news/2001/011123b.htm

Dutch Ministry for Nature Management (1997). Timber certification and sustainable forestry. Department of Nature Management, Dutch Ministry of Agriculture, Nature Management and Fisheries, Feb 1997.

DWW (1994). *Wijzer 65, Hout in de waterbouw.* [Timber in the waterways – in Dutch]. [Rijkswaterstaat] Dienst Weg- en Waterbouwkunde, Delft, The Netherlands.

ECOPRO (1996). *Environmentally friendly coastal protection, Code of Practice.* Eolas Offshore & Coastal Engineering Unit, Dublin, Ireland.

Environment Agency (1999). *Waterway bank protection: a guide to erosion assessment and management.* Environment Agency, Bristol, UK.

Environment Agency (2003). UK Climate Impacts Programme 2002 Climate Change Scenarios: implementation for flood and coastal defence: guidance for users. Environment Agency R&D Technical Report W5B-029/TR. Available from http://www.environment-agency.gov.uk/commondata/105385/w5b_029_tr.pdf

Environmental Audit Committee (2002). *Buying time for forests: Timber trade and public procurement.* The Stationery Office. Available from http://www.parliament.the-stationery-office.co.uk/pa/cm200102/cmselect/cmenvaud/792/792.pdf

Environmental Impact Assessment (EIA) Regulations (1999). *The town and country planning (Environmental Impact Assessment (England and Wales) Regulations.* Statutory instrument 1999 No. 293. Available from http://www.hmso.gov.uk/si/si1999/19990293.htm

Environmental Resources Management (ERM) (2002). Procurement of timber products from 'legal & sustainable sources' by Government & its executive agencies : Scoping Study Report. Available from www.forestforum.org.uk/docs/scopingstudyreport6-8-02.pdf

EU FLEGT (2002). *The EU & Forest Law Enforcement, Governance and Trade.* Available from http://europa.eu.int/comm/external_relations/flegt/intro/

European Forestry Institute (EFI) (2000). *Criteria and indicators for sustainable management at the forest management unit level.* Available from http://efi.fi/publications/Proceedings/38.html

FAO (2001). *Global Forest Resources Assessment 2000.* United Nations Food and Agriculture Organization, Rome. Available from http://www.fao.org/forestry/fo/fra/main/index.jsp

Fleming, C.A. (1990). Guide on the uses of groynes in coastal engineering. Construction Industry Research and Information Association, Report 119, London.

G8 (1998). G8 Action Programme on Forests. Available from http://birmingham.g8summit.gov.uk/forfin/forests.shtml

Gardner, J.D., Townend, I.H. and Fleming, C.A. (1986). The design of a slotted vertical screen breakwater. *Proc. 20th Intrnl Conf. on Coastal Engineering,* American Society of Civil Engineers.

Graham, J.S. (1983). Design of pressure-treated wood bulkheads. *Proc. Coastal Structures '83,* pp. 286-94, American Society of Civil Engineers.

Guyana Forestry Commission (2002). *Guyana timber grading rules for hardwoods.* 3rd Ed. produced with support from the ACP-EU Trade development project. Pavnik Press.

Health and Safety Executive (2003). Toxic woods: Wood working sheet 30.

HMSO (1997). *Handbook of hardwoods.* Her Majesty's Stationary Office, London.

House of Commons (2002). Buying Time for Forests: Timber Trade and Public Procurement. Environmental Audit Committee. House of Commons. Available from http://www.parliament.the-stationery-office.co.uk/pa/cm200102/cmselect/cmenvaud/792/79202.htm

IWEM (1987). *River engineering – Part I, Design principles.* Brandon, T.W. (ed.) The Institution of Water and Environmental Management, London.

IWEM (1989). *River engineering – Part II, Structures and coast defence works.* Brandon, T.W. (ed.) The Institution of Water and Environmental Management, London.

MAFF (1999). Flood and coastal defence project appraisal guidance: Economic Appraisal. Ministry of Agriculture Fisheries and Food (now Department for the Environment, Food and Rural Affairs) FCDPAG3, Dec.

MAFF (2000a). Flood and coastal defence project appraisal guidance: Approaches to risk. Ministry of Agriculture Fisheries and Food (now Department for the Environment, Food and Rural Affairs) FCDPAG4, Feb.

MAFF (2000b). Flood and coastal defence project appraisal guidance: Environmental Appraisal. Ministry of Agriculture Fisheries and Food, (now Department for the Environment, Food and Rural Affairs) FCDPAG5, March.

MAFF (2001a). Flood and coastal defence project appraisal guidance: Strategic planning and appraisal. Ministry of Agriculture Fisheries and Food (now Department for the Environment, Food and Rural Affairs) FCDPAG2, April.

MAFF (2001b). Flood and coastal defence project appraisal guidance: Overview (including general guidance). Ministry of Agriculture Fisheries and Food, (now Department for the Environment, Food and Rural Affairs) FCDPAG1, May.

Masters, N. (2001). *Sustainable use of new and recycled materials in coastal and fluvial construction – A guidance manual.* Thomas Telford, London.

McBride, M.W., Smallman, J.V. and Allsop, N.W.H. (1996). Guidelines for the hydraulic design of harbour entrances. HR Wallingford Report SR 430, Feb.

Morris, M.W. and Simm, J.D. (2000). *Construction risk in river and estuary engineering – A guidance manual.* Thomas Telford, London.

Ozelton, E.C. and Baird, J.A. (2002). *Timber Designers' Manual.* Blackwell Science, Oxford.

PIANC (1990). Inspection, maintenance and repair of maritime structures exposed to material degradation caused by a salt water environment. Permanent International Association of Navigational Congresses, Report of Working Group No. 17 of the Permanent Technical Committee II.

RIIA (2002). Controlling the trade in illegal logging. Royal Institute of International Affairs, London. Available from http://www.riia.org/pdf/briefing_papers/tradeinillegaltimber.pdf

Rio (1992). Non-legally binding statement of forest principles. United Nations. Available from http://www.un.org/documents/ga/conf151/aconf15126-3annex3.htm

SALVO (2003). Website for architectural salvage, garden antiques and reclaimed building materials. Available from http://www.salvo.co.uk/index.html

Schiereck, G.J. (2001). *Introduction to bed, bank and shore protection: engineering the interface of soil and water.* Delft University Press, The Netherlands.

Scottish Natural Heritage (SNH) (2002). A guide to managing coastal erosion in beach/dune systems. Available from http://www.snh.org.uk/pdfs/publics/catalogue/nhm.pdf

Searle, S.A. (1975). *The tidal threat.* Dunes Group, UK.

Simm, J.D. and Cruickshank, I.C. (1998). *Construction risk in coastal engineering.* Thomas Telford, London.

Simm, J.D. and Masters, N.D. (2003). *Whole life costs and project procurement in port, coastal and fluvial engineering – how to escape the costs boxes.* Thomas Telford, London.

Town and Country Planning Regulations 1999. The Town and Country Planning (Trees) Regulations 1999.

TRADA Technology (1999). *Timber decking manual.* Timber Research and Development Association, High Wycombe, UK.

Upton, C. and Bass, S. (1995). *The forest certification handbook.* Earthscan Publications, London.

USACE (1992). Coastal groins and nearshore breakwaters. US Army Corps of Engineers, Engineer Manual 1110-2-1617, Aug.

USACE (2002). Coastal engineering manual. EM 1110-2-1100. Draft – September 2002. Available from http://bigfoot.wes.army.mil/cem026.html.

WWF (2002a). *Forests facts and key issues.* World Wildlife Fund. Available from http://www.wwf.org.uk/researcher/programmethemes/forests/0000000196.asp

WWF (2002b). *The timber footprint of the G8 and China.* World Wildlife Fund. Available from http://www.wwf-uk.org/filelibrary/pdf/g8timberfootprint.pdf

12. Bibliography

Altiero, K. (1997). Engineers ignore significant deterioration. *Civil Engineering – ASCE*, Vol. 67, No.1, Jan 1997, pp. 30-1.

ASCE (1992). Clean Waters taking a toll on timber structures. *Civil Engineering – ASCE*, Vol. 62, No.3, March 1992, pp. 28-9.

Baileys, R.T. (1995). Timber structures and the marine environment. *Proc. Ports '95*, pp. 703-10, ASCE.

Brampton, A.H. (2002). *ICE Design and practice guide – Coastal defence.* Thomas Telford, London.

Brampton, A.H. and Motyka, J.M. (1983). The effectiveness of groynes. *Proc. Shoreline Protection*, pp. 151-56, Thomas Telford, London.

Brazier, J.D. (1996). *A review of UK and European consumption of tropical hardwoods.* Building Research Establishment/Construction Research Communications, London.

BRE (1996). *Hardwoods for construction and joinery: current and future sources of supply.* Building Research Establishment, Digest 417, Aug 1996.

BS EN 392: 1995 *Glued laminated timber – Shear test of glue lines.*

BTCV (2003a). *Sand Dunes.* British Trust for Conservation Volunteers. Available from http://handbooks.btcv.org.uk/pubs/home

BTCV (2003b). *Waterways & Wetlands.* British Trust for Conservation Volunteers. Available from http://handbooks.btcv.org.uk/pubs/home

Buslov, V.M. and Scola, P.T. (1991). Inspection and structural evaluation of timber pier: case study. *Structural Engineering – ASCE*, Vol. 117, No. 9, Sept 1991.

Chen, S. and Kim, R. (1996). Condition assessment of marine timber piles using stress wave method. *Proc. Structures Congress*, Chicago, April 1996, American Society of Civil Engineers.

Coffer, W.F., McLean, D.I. and Wolcott, M.P. (2001). Structural evaluation of engineered wood composites for naval waterfront facilities. *Proc. Ports '01*, Section 3 Chapter 1, American Society of Civil Engineers.

DD 239: 1998 *Recommendations for preservation of timber.*

Defra (2002). *Procurement of timber products from legal & sustainable sources by Government & its executive agencies.* Department for Environment, Fisheries and Rural Affairs. Available from http://www.forestforum.org.uk/tradeb.htm

Eaton, R.A. and Hale, M.D.C. (1993). *Wood. Decay, pests and protection.* Chapman and Hall, London.

FoE (2002). *Good wood guide.* Friends of the Earth, Flora & Fauna International, London.

FPL (1999). Wood Handbook: Wood as an engineering material. Forest Products Laboratory General Technical Report FPL-GTR-113. Available from http://www.fpl.fs.fed.us

FSC (2000). FSC Principles and Criteria. Forestry Stewardship Council document 1.2 revised Feb 2000. Available from: http://www.fscoax.org/html/1-2.html

FSC (2003). FSC Principles and Criteria. Forestry Stewardship Council. Available from http://www.fscoax.org/principal.htm

Graham, J.S. (1983). Design of pressure-treated wood bulkheads. *Proc. Coastal Structures '83*, pp. 286-94, American Society of Civil Engineers.

Hester, J.D.S. (1983). Timber in shoreline protection. *Proc. Shoreline Protection*, pp. 199-202, Thomas Telford, London.

Kemp, P. (1995). Tropical hardwoods in the maritime environment – a case for their continued use. Proceedings of a half day meeting held in Oct 1995, Institution of Civil Engineers, London.

Kermani, A. (1999). *Structural timber design.* Blackwell Science, Oxford.

Mettem, C.J. and Richens, A.D. (1991). *Hardwoods in construction.* Timber Research and Development Association, TBL 62.

Metzger, S.G. and Abood, K.A. (1998). The Limnoria has Landed! *Proc. Ports '98.* pp. 672-81. American Society of Civil Engineers.

NAVFAC (2001). Maintenance of waterfront facilities. US Naval Facilities Engineering Command UFC 4-150-07, June.

NRA (1993). Specification and use of timber for marine and estuarine construction. National Rivers Authority R&D Note 133.

NRA (1994). Timber treatment chemicals: Priorities for environmental quality standard development. National Rivers Authority Note 340.

Nussbaum, R., Jennings, S. and Garforth, M. (2002). *Assessing forest certification schemes: a practical guide.* Oxford: Proforest for Department for International Development. Available from http://www.proforest.net/objects/gscheme2.pdf

Oliver, A.C. (1974). Timber for marine and fresh water construction. Timber Research and Development Association, Dec.

Oliver, A.C. and Woods, R.P. (1959). The resistance of certain timbers in sea defence groynes to shingle abrasion. Timber Research and Development Association, B/TR/4, Dec.

Simm, J.D. and Camilleri, A.J. (2001). Construction risk in coastal and river engineering. *Chartered Institute of Water & Environmental Management*, Vol. 15, Nov, pp. 258-64.

Statutory Instrument 1999 No. 1892. Available from http://www.hmso.gov.uk/si/si1999/19991892.htm

Steer, P.J. (2001). EN 1995 Eurocode 5: Design of timber structures. *Civil Engineering – ICE*, Vol. 144, Nov. 2001, pp. 39-43.

Tanal, V. and Matlin, A. (1996). Marine borers are back. *Civil Engineering – ASCE*, Vol. 66, No. 10, Oct. 1996, pp. 71-3.

Taylor, R.B. (1995). Composite recycled plastic marine piling and timber: An alternative to traditional wood products for marine use. *Proc. Ports '95*, pp. 711-22, American Society of Civil Engineers.

Thomas, R.S. and Hall, B. (1992). *Seawall design.* Construction Industry Research Association, London.

TRADA (1991a). Wood preservation – processing and site control. Timber Research and Development Association, Wood information sheet 2/3 – 16.

TRADA (1991b). Wood preservation – a general background: Part 1 – The risks. Timber Research and Development Association, Wood information sheet 2/3 – 32.

TRADA (1991c). Wood preservation – a general background: Part 2 – Chemicals and processes. Timber Research and Development Association, Wood information sheet 2/3 – 33.

TRADA (1995a). Durability and preservative treatment of wood – European Standards. Timber Research and Development Association, Wood information sheet 2/3 – 38.

TRADA (1995b). Durability and preservative treatment – Key British and European Standards. Timber Research and Development Association, Wood information sheet 2/3 – 39.

TRADA (1995c). Preservative treated wood – European Standards – Test methods. Timber Research and Development Association, Wood information sheet 2/3 – 40.

TRADA (1999). Timbers – their properties and uses. Timber Research and Development Association, Wood information sheet 2/3 –10.

TTJ (2002). Moves to limit CCA in US as well as EU. *Timber Trades Journal.* 16 February 2002. Available from http://www.worldwidewood.com/archive/news/nfeb02/16cca.htm.

USACE (1984). Shore protection manual (Vol. I and II). 4th Edn, USACE Coastal Engineering Research Center.

Webber, D. and Yao, J. (2001). Effectiveness of pile wraps for timber bearing piles. Proc. Ports '01, Section 22, Chapter 1, American Society of Civil Engineers.

WSU (2002). Program Overview: Naval Advanced Wood Composites. Available from http://www.composites.wsu.edu/navy/Navy2/main.ProgramOver.html

Appendices

Appendix 1.
Species properties and characteristics

This appendix details a number of timber species that are considered suitable for use in the marine and/or fresh water environment. The list is by no means exhaustive. No attempt has been made to identify those species which are known to have credible certification demonstrating good environmental management as this situation is constantly changing.

From a technical standpoint, many of the timbers detailed in this appendix meet the requirements for use in the marine and/or fresh water environments. However, in addition to the material properties of the timber, important commercial specification requirements may influence the choice of timber species.

These include volumes, sizes, ease of procurement, continuity of supply and price. Commercial requirements have tended, in the past, to influence the choice of timber available to the marine engineer. We have made no attempt to rank the commercial availability of the listed timbers as it is likely that their availability will change over the course of time although general comments on ease of sourcing have been made where information is available. Some species may become harder to procure whilst others may gain in market acceptance.

With reference to marine borer resistance, the results for species that have been tested as part of the investigative laboratory screening programme have been marked with an asterix (*) and presented as a footnote to the text.

With reference to natural durability, as far as can be determined, the ratings have been assigned on the basis of 'graveyard' tests described in the *Handbook of hardwoods* (HMSO, 1997). Other references are cited where appropriate.

The sizes given in the species tables are those in common use in marine and fresh water construction. In some cases the sizes are given as produced in the exporting nation. Some exporting nations, e.g. Guyana, still produce their timber to imperial dimensions and, furthermore, control their sizes within the limits of tolerances laid down in their national grading rules.

It is imperative that the designer/specifier is particularly diligent when procuring timber and consults their timber supplier at the earliest possible opportunity to determine the actual sizes that are available from the producer nation. The route for timber procurement should, ideally, be investigated at an early stage, so that the timber importer is made aware of the sizes and volumes required by the designer. This will enable the importer to establish whether the shipper can supply the required volumes and sizes of the specified species or species mix. Where sizes are not given in the text, the tree characteristics, i.e. bole length, etc. may provide an indication as to sizes that may be obtainable.

Mechanical strength data, where available, have been included for each species. These data may be found in a number of documents and data for a particular species may vary between documents. This is generally a reflection of the amount of material used for testing and also demonstrates the natural variability of timber as a structural material. The strength values given in this appendix are primarily for the comparison of species and are taken from the publications listed below. (Where appropriate, the reference identifying the source for each value is given). The values given are average ultimate strength values based on testing small, clear defect-free sections of the candidate species. Each value is subject to some degree of error and small differences between the mean values for species may exist. It is important to stress that these values are not design values based on strength classes such as those found in BS 5268-2: 2002. The data may be used in the derivation of design stresses although, with the exception of the method described in Section 4.3.3 of this report, precise methods for derivation of design stresses from ultimate strength values are beyond the scope of this document.

Individual references are cited where key attributes such as natural durability, density and marine borer resistance are presented. Other attributes such as the working properties of the timber, uses and the tree characteristics have been taken from a number of relevant publications and no attempt has been made to assign each attribute to its relevant source. The sources of information are cited below. Where no information is given for certain attributes it is because no information was available from the references cited below, and in some instances is a reflection of the lack of readily available technical data for lesser known species of timber that are, currently, of little commercial importance outside their countries of origin. The references used to obtain the material properties of the timbers listed is not exhaustive, however, and information may probably be obtained by a more detailed literature search.

References for timber species data

1. *Handbook of hardwoods* (HMSO, 1997)
2. *Timber for marine and fresh water construction* (TRADA, 1974)
3. *The strength properties of timber* (HMSO, 1983)
4. *Ecotimber handbook* (Ecotimber, 2001-2003)
5. *Timbers of the world: Volumes 1-9* (TRADA, 1978)
6. *Characteristics, properties and uses of timber. Vol. 1. South-East Asia, Northern Australia and the Pacific* (CSIRO*, 1982)
7. *South American timbers* (CSIRO, 1979)
8. *Handbook of softwoods* (HMSO, 1972)

*Commonwealth Scientific and Industrial Research Organisation.

Abiurana	**Pouteria guianensis**		
Other names:	Abiurana-branca, Abiurana-abiu		
Source of supply:	South and Central America		
Available sizes:	Length Width Thickness		
Mechanical properties:		*Green*	*Dry*
	Bending strength (N/mm^2) Modulus of Elasticity (kN/mm^2) Compression strength parallel to grain (N/mm^2) Maximum shear strength parallel to grain (N/mm^2) Impact strength (m)		
Source:	Ecotimber handbook		
Density:	kg/m^3		900
Movement:			
Natural durability:	Very durable		
Marine borer resistance:	Not resistant *		
Treatability of sapwood:	Probably resistant		
Treatability of heartwood:	Probably resistant		
Source:	Ecotimber handbook		
Tree characteristics, working properties and uses:	Used for heavy construction and hydraulic and marine construction.		
Availability:	By forward shipment. No information on sizes or volumes available.		

*Performed poorly in laboratory tests

Acaria quara	**Minquartia guianensis**		
Other names:	Manwood, Macaa, Pachiche		
Source of supply:	South America		
Available sizes:	Length Width Thickness		
Mechanical properties:		*Green*	*Dry*
	Bending strength (N/mm^2) Modulus of Elasticity (kN/mm^2) Compression strength parallel to grain (N/mm^2) Maximum shear strength parallel to grain (N/mm^2) Impact strength (m)		
Source:			
Density:	kg/m^3		700
Movement:	Large		
Natural durability: *Marine borer resistance:* *Treatability of sapwood:* *Treatability of heartwood:*	Very durable Probably resistant * Probably resistant Probably resistant		
Source:	Ecotimber handbook		
Tree characteristics, working properties and uses:	The tree reaches a height of 30-40 m with a bole length of 10-20 m. The diameter of the bole can reach up to 1.0m. The bole is reported to be cylindrical or buttressed.		
Availability:	By forward shipment. No information on sizes or volumes available.		

* Only tested against Limnoria spp. to-date

Afzelia	**Afzelia spp.**		
Other names:	Apa, Aligna (Nigeria), Doussie (Cameroon)		
Source of supply:	West and Central Africa		
Available sizes:	Length Width Thickness		

Mechanical properties:		*Green*	*Dry*
	Bending strength (N/mm^2)	89	97
	Modulus of Elasticity (kN/mm^2)	8.7	8.5
	Compression strength parallel to grain (N/mm^2)	47	71
	Maximum shear strength parallel to grain (N/mm^2)	13.4	19.7
	Impact strength (m)	1.04	0.71

Source:

Density:	kg/m^3	1137	685

Movement: Small

Natural durability: Very durable
Marine borer resistance: Probably resistant
Treatability of sapwood: Moderately resistant
Treatability of heartwood: Extremely resistant

Source: Handbook of hardwoods (HMSO, 1997)

Tree characteristics, working properties and uses: A heavy timber with good strength properties and excellent natural durability. Interlocked grain – difficult to work. The tree may reach a height in excess of 30 m. The bole is fairly cylindrical and 1.0 m in diameter.

Availability: Limited

Amiemfo-samina **Albizia ferruginea**

Other names: West African Albizia, Okuro, Ayinre, Sifou

Source of supply: West and Central Africa

Available sizes: Length
Width
Thickness

Mechanical properties:

	Green	Dry
Bending strength (N/mm^2)		105
Modulus of Elasticity (kN/mm^2)		10.7
Compression strength parallel to grain (N/mm^2)		64.7
Maximum shear strength parallel to grain (N/mm^2)		16.6
Impact strength (m)		0.56

Source: Handbook of hardwoods (HMSO, 1997) and The strength properties of timber (HMSO, 1983)

Density: kg/m^3 705

Movement: Small

Natural durability: Very durable
Marine borer resistance: Probably resistant *
Treatability of sapwood: Permeable
Treatability of heartwood: Extremely resistant

Source: Handbook of hardwoods (HMSO, 1997)

Tree characteristics, working properties and uses: Interlocked grain – difficult to work. The dust may cause irritation. The tree may reach a height in excess of 37 m. Straight bole, length 9-12 m and 1.0 m in diameter. Used for marine piling and heavy construction, flooring, furniture and plywood.

Availability: No information available

*Performed better than greenheart when exposed to Limnoria spp. during laboratory trials

Andaman padauk | **Pterocarpus dalbergioides**

Other names: Kokrodua

Source of supply: Andaman Isles

Available sizes: Length
Width
Thickness

Mechanical properties:

	Green	Dry
Bending strength (N/mm^2)	88	105
Modulus of Elasticity (kN/mm^2)	10.3	11.2
Compression strength parallel to grain (N/mm^2)	48.8	61.5
Maximum shear strength parallel to grain (N/mm^2)		
Impact strength (m)		

Source:

Density: kg/m^3 | 900 | 770

Movement: n/a

Natural durability: Very durable
Marine borer resistance: Probably resistant
Treatability of sapwood: Permeable
Treatability of heartwood: Moderately resistant

Source: Handbook of hardwoods (HMSO, 1997)

Tree characteristics, working properties and uses: Interlocked grain – difficult to work. Slight tendency to split. Little distortion during drying. The timber has high resistance to abrasion. Little reported use as pilings. Suitable for heavy constructions. The tree is reported to reach a height of 25-37 m with a diameter of 0.75-0.9 m. The bole is straight and cylindrical.

Availability: No information available

Andira	**Andira spp.**		
Other names:	Sucupira vermelho, Almendro, Rode kabbes		
Source of supply:	South America		
Available sizes:	Length Width Thickness		
Mechanical properties:		*Green*	*Dry*
	Bending strength (N/mm^2)		143
	Modulus of Elasticity (kN/mm^2)		16.3
	Compression strength parallel to grain (N/mm^2)		74
	Maximum shear strength parallel to grain (N/mm^2)		
	Impact strength (m)		
Source:			
Density:	kg/m^3	1200	685
Movement:	Small		
Natural durability:	Very durable		
Marine borer resistance:	Probably resistant		
Treatability of sapwood:	Probably resistant		
Treatability of heartwood:	Probably resistant		
Source:	Ecotimber handbook		
Tree characteristics, working properties and uses:	The tree reaches a height of 40-60 m with a bole length of 20-30 m.		
Availability:	By forward shipment. No information on sizes or volumes available.		

Angelim vermelho   **Hymenolobium spp.**

Other names: Angelim pedra, Sapupira amarella, Amgelim de mata

Source of supply: Brazil, Guianas and Venezuela

Available sizes: Length
Width
Thickness

Mechanical properties:

	Green	Dry
Bending strength (N/mm^2)		131
Modulus of Elasticity (kN/mm^2)		16.4
Compression strength parallel to grain (N/mm^2)		66
Maximum shear strength parallel to grain (N/mm^2)		
Impact strength (m)		

Source:

Density: kg/m^3 650-800

Movement: Large

Natural durability: Very durable
Marine borer resistance: Moderately resistant *
Treatability of sapwood: Moderately resistant
Treatability of heartwood: Extremely resistant

Source: Ecotimber handbook

Tree characteristics, working properties and uses: The tree reaches a height of 40-60 m with a bole length of 20-30 m. The diameter of the bole can reach up to 3.5 m. The bole is cylindrical and buttressed. The timber is reported to be suitable for marine constructions.

Availability: By forward shipment. No information on sizes or volumes available.

*Performed better than greenheart when exposed to Limnoria spp. during laboratory trials

Ayan **Distemonanthus benthamianus**

Other names: Movingui (France), Ayanran (Nigeria), Bonsamdua (Ghana), Distemonanthus (Great Britain)

Source of supply: West Africa

Available sizes: Length
Width
Thickness

Mechanical properties:

	Green	Dry
Bending strength (N/mm^2)		108
Modulus of Elasticity (kN/mm^2)		15.7
Compression strength parallel to grain (N/mm^2)		57.3
Maximum shear strength parallel to grain (N/mm^2)		14.5
Impact strength (m)		0.84

Source:

Density: kg/m^3 600-770

Movement: Small

Natural durability: Moderately durable
Marine borer resistance: Probably resistant *
Treatability of sapwood: Permeable
Treatability of heartwood: Extremely resistant

Source: Handbook of hardwoods (HMSO, 1997) and The strength properties of timber (HMSO, 1983)

Tree characteristics, working properties and uses: A moderately heavy timber with good dimensional stability and may be used as an oak substitute. The tree is reported to reach a height of 27-38 m with a diameter of 0.8-1.4 m. The bole is reasonably straight and cylindrical.

Availability: No information on availability in the UK

* Only tested against Limnoria spp. to-date and good performance observed

Balau	**Shorea spp. including S. glauca, S. ciliata, S. gisok, S. laevifolia**
Other names:	Selangan batu. The names balau and selangan batu describe a range of heavy Shores spp. timbers. It is important to recognise that when the prefix 'red' is used, this refers to balau of lesser durability. For marine constructions, only selangan batu or yellow balau should be considered. The variation in mechanical properties reflects the variability of balau as balau refers to a wide range of species.
Source of supply:	Sabah (Malaysia)

Available sizes:

Length	Up to 9 000 mm
Width	100-350 mm
Thickness	50-350 mm

Mechanical properties:

	Green	Dry
Bending strength (N/mm^2)	96-11	122-143
Modulus of Elasticity (kN/mm^2)	11.7-14.3	13.7-17.1
Compression strength parallel to grain (N/mm^2)	50.1-58.3	65.6-77.2
Maximum shear strength parallel to grain (N/mm^2)	11.6	15.5
Impact strength (m)	0.94-1.14	1.02-1.35

Source:

Density:	kg/m^3	990-1150	750-900

Movement:

Natural durability:	Very durable/durable
Marine borer resistance:	Moderately resistant
Treatability of sapwood:	Probably resistant
Treatability of heartwood:	Extremely resistant

Source:	Timber for marine and fresh water construction (TRADA, 1974)
Tree characteristics, working properties and uses:	Used in all heavy constructions. Grain – difficult to work. Slight tendency to split. Little distortion during drying. The timber has high resistance to abrasion. Little reported use as pilings. Suitable for heavy constructions and sea defences. Section sizes larger than 305 mm × 305 mm are uncommon and require pre-ordering; lengths over 7.5 m are limited.
Availability:	Readily available

Basralocus	**Dicoryia guianensis**	
Other names:	Angelique	
Source of supply:	Surinam, French Guiana, Guyana, Brazil	

Available sizes:	Length	Up to 21 300 mm
	Width	225-550 mm
	Thickness	225-550 mm

Mechanical properties:		Green	Dry
	Bending strength (N/mm^2)	82	
	Modulus of Elasticity (kN/mm^2)	12.5	
	Compression strength parallel to grain (N/mm^2)	41.2	
	Maximum shear strength parallel to grain (N/mm^2)	10.5	
	Impact strength (m)		

Source:	Timber for marine and fresh water construction (TRADA, 1974)		
Density:	kg/m^3	960-1072	710-900

Movement:

Natural durability:	Very durable
Marine borer resistance:	Resistant
Treatability of sapwood:	Probably resistant
Treatability of heartwood:	Extremely resistant

Source:	Handbook of hardwoods (HMSO, 1997)

Tree characteristics, working properties and uses: Difficult to work owing to high silica content (this confers resistance to marine borers). The timber has tendency to split. Proven track record for use in marine constructions. Suitable for all heavy construction purposes. High abrasion resistance. The tree reaches a height of 30 m or more with a diameter of up to 1.5 m.

Availability: Readily available from specialist importers. Sizes usually confirmed by negotiation with shipper.

Belian	**Eusideroxylon zwageri**

Other names: Borneo ironwood, Billiqan, Onglen, Ijzerhout

Source of supply: The species occurs throughout Borneo (Sabah, Sarawak and Kalimantan). The trade names reflect the state of origin and colonial history, i.e. Ijzerhout is from Indonesian Kalimantan, formerly part of the Dutch East Indies.

Available sizes:

Length	Up to 7 600 mm
Width	75-300 mm
Thickness	50-300 mm

Mechanical properties:

	Green	*Dry*
Bending strength (N/mm^2)	143	178
Modulus of Elasticity (kN/mm^2)	17.7	18.3
Compression strength parallel to grain (N/mm^2)	79.9	93.9
Maximum shear strength parallel to grain (N/mm^2)	15.4	20.3
Impact strength (m)	1.17	1.09

Source: The strength properties of timber (HMSO, 1983)

Density: kg/m^3 1025

Movement:

Natural durability:	Very durable
Marine borer resistance:	Resistant
Treatability of sapwood:	Resistant
Treatability of heartwood:	Resistant

Source: Timber for marine and fresh water construction (TRADA, 1974)

Tree characteristics, working properties and uses: Difficult to work. Suitable for marine constructions. High abrasion resistance.

Availability: Extremely limited. The timber is rarely specified in the UK, therefore there is little incentive for Malaysia or Indonesia to export belian into the UK market place.

Bompagya	**Mammea africana**		
Other names:	African Apple, Oboto		
Source of supply:	West and Central Africa		
Available sizes:	Length Width Thickness		
Mechanical properties:		*Green*	*Dry*
	Bending strength (N/mm^2) Modulus of Elasticity (kN/mm^2) Compression strength parallel to grain (N/mm^2) Maximum shear strength parallel to grain (N/mm^2) Impact strength (m)		
Source:			
Density:	kg/m^3		690
Movement:			
Natural durability:	Very durable		
Marine borer resistance:	Probably resistant *		
Treatability of sapwood:	Probably resistant		
Treatability of heartwood:	Resistant		
Source:	Timbers of the world (TRADA)		
Tree characteristics, working properties and uses:	The tree can be up to 36 m high and 3.6 m in girth. Used for railway sleepers, exterior joinery and carpentry, and heavy construction work.		
Availability:	No information on availability in the UK		

*Similar performance to greenheart in preliminary screening trials against Limnoria spp.

Bruguiera   **Bruguiera gymnorrhiza and B. parviflora**

Other names: Black mangrove

Source of supply: Malaysia, Papua New Guinea

Available sizes: Length
Width
Thickness

Mechanical properties:

	Green	Dry
Bending strength (N/mm^2)	73-86	114-134
Modulus of Elasticity (kN/mm^2)		14.2-16.3
Compression strength parallel to grain (N/mm^2)		62-71
Maximum shear strength parallel to grain (N/mm^2)		15-16.7
Impact strength (m)		

Source:

Density: kg/m^3 805-1010

Movement:

Natural durability: Moderately durable
Marine borer resistance: Very resistant *
Treatability of sapwood: Permeable
Treatability of heartwood: Fairly easy

Source: Characteristics, properties and uses of timber. Vol 1. South-East Asia, Northern Australia and the Pacific. CSIRO

Tree characteristics, working properties and uses: Timber is hard and strong and works fairly well. The tree may reach 15-30 m in height with a straight bole up to 15 m and 0.3-0.8 m diameter.

Availability:

*Excellent resistance to Limnoria spp. observed during laboratory screening trials

Brush box	**Tristania conferata**		
Other names:	Pink box		
Source of supply:	Eastern Australia		
Available sizes:	Length Width Thickness		
Mechanical properties:		*Green*	*Dry*
	Bending strength (N/mm²)		
	Modulus of Elasticity (kN/mm²)		
	Compression strength parallel to grain (N/mm²)		
	Maximum shear strength parallel to grain (N/mm²)		
	Impact strength (m)		
Source:			
Density:	kg/m³	1080	900
Movement:			
Natural durability:	Moderately durable		
Marine borer resistance:	Probably resistant as siliceous		
Treatability of sapwood:	Resistant		
Treatability of heartwood:	Resistant		
Source:	Handbook of hardwoods (HMSO, 1997)		
Tree characteristics, working properties and uses:	Difficult to work, tendency to distort upon drying. High strength properties with good resistance to abrasion. High silica content would suggest that this species is resistant to marine borers. A large tree reported to reach a height of 36 m with a diameter of up to 2.0 m.		
Availability:	Extremely limited		

Chengal	**Balanocarpus heimii**	
Other names:		
Source of supply:	Malaysia and South East Asia	
Available sizes:	Length	Up to 9 000 mm
	Width	150-300 mm
	Thickness	100-300 mm

Mechanical properties:		*Green*	*Dry*
	Bending strength (N/mm^2)	12.8	n/a
	Modulus of Elasticity (kN/mm^2)	18.1	
	Compression strength parallel to grain (N/mm^2)	71.5	
	Maximum shear strength parallel to grain (N/mm^2)	15.4	
	Impact strength (m)		

Source: Timber for marine and fresh water construction (TRADA, 1974)

Density:	kg/m^3	1040-1120	836

Movement: Small

Natural durability: Very durable
Marine borer resistance: Unknown
Treatability of sapwood: Resistant
Treatability of heartwood: Resistant

Source: Timber for marine and fresh water construction (TRADA, 1974)

Tree characteristics, working properties and uses: Dense, hard timber with strength properties reportedly similar to greenheart. The timber has interlocked grain and is reported to dry easily and not suffer from distortion. The timber is also reported to be stable in service. Chengal is suitable for all types of heavy constructions and is easy to work.

Availability: Extremely limited. The timber is rarely specified in the UK, therefore there is little incentive for Malaysia to export chengal into the UK market place.

Courbaril	**Hymeneae spp.**		
Other names:	Jatoba, Jatai vermelho, Jata armarelo, Farinheira		
Source of supply:	Brazil and Central America		
Available sizes:	Length Width Thickness		
Mechanical properties:		*Green*	*Dry*
	Bending (N/mm^2)		
	Modulus of Elasticity (kN/mm^2)		
	Compression strength parallel to grain (N/mm^2)		
	Maximum shear strength parallel to grain (N/mm^2)		
	Impact strength (m)		
Source:			
Density:	kg/m^3		700-900
Movement:	Medium		
Natural durability:	Moderately durable		
Marine borer resistance:	Probably resistant *		
Treatability of sapwood:	Resistant		
Treatability of heartwood:	Resistant		
Source:	Timbers of the world (TRADA)		
Tree characteristics, working properties and uses:	The tree grows to a height of 40 m with a diameter ranging from 0.6 m to 1.2 m. The bole varies in length from 12 m to 24 m and is buttressed. Courbaril is a very strong, hard, tough timber that is generally superior to oak. It has been reported as having been used for heavy construction in Brazil.		
Availability:	Readily available		

*Performed better than greenheart when exposed to Limnoria spp. during laboratory trials

Cumaru	**Dipterix spp.**		
Other names:			
Source of supply:	South America		
Available sizes:	Length Width Thickness		

Mechanical properties:		*Green*	*Dry*
	Bending strength (N/mm^2)		199
	Modulus of Elasticity (kN/mm^2)		22.0
	Compression strength parallel to grain (N/mm^2)		105
	Maximum shear strength parallel to grain (N/mm^2)		
	Impact strength (m)		

Source:	Ecotimber handbook		
Density:	kg/m^3	1200	1070
Movement:	Medium		

Natural durability:	Very durable
Marine borer resistance:	Probably resistant *
Treatability of sapwood:	Probably resistant
Treatability of heartwood:	Probably resistant

Source:	Ecotimber handbook

Tree characteristics, working properties and uses: The tree reaches a height of 40-60 m with a bole length of 10-30 m. The diameter of the bole can reach up to 1.5 m. The bole is buttressed.

Availability: By forward shipment. No information on sizes or volumes available.

*Performed better than greenheart in laboratory screening trials against Limnoria spp.

Cupiuba   **Goupia glabra**

Other names: Kabukalli, Copi, Goupi

Source of supply: Brazil, Guyana

Available sizes: Length
                     Width
                     Thickness

Mechanical properties:

	Green	Dry
Bending strength (N/mm^2)		122
Modulus of Elasticity (kN/mm^2)		14.7
Compression strength parallel to grain (N/mm^2)		62
Maximum shear strength parallel to grain (N/mm^2)		
Impact strength (m)		

Source: Ecotimber handbook and South American timbers, 1979 CSIRO

Density: kg/m^3 1130   840

Movement: Medium

Natural durability: Very durable
Marine borer resistance: Probably resistant *
Treatability of sapwood: Probably resistant
Treatability of heartwood: Probably resistant

Source: Timbers of the world (TRADA)

Tree characteristics, working properties and uses: The tree grows to about 37 m with a long trunk. Timber is very heavy, but works fairly well. It is used for bridge decking, railway sleepers, carpentry and joinery.

Availability: Regular

*Excellent resistance to Limnoria spp. observed during laboratory screening trials

Dahoma	**Piptadeniastrum africanum**		
Other names:	Agboin, Dabema, Ekhimi		
Source of supply:	West, central and some parts of East Africa		

Available sizes:	Length	Up to 7 600 mm
	Width	Up to 450 mm
	Thickness	Up to 450 mm

Mechanical properties:		*Green*	*Dry*
	Bending strength (N/mm^2)	76	109
	Modulus of Elasticity (kN/mm^2)	9.9	11.2
	Compression strength parallel to grain (N/mm^2)	36.7	58.7
	Maximum shear strength parallel to grain (N/mm^2)	11.2	17.6
	Impact strength (m)	0.89	0.84

Source:	The strength properties of timber (HMSO, 1983)		
Density:	kg/m^3	849	689
Movement:	Medium		

Natural durability:	Moderately durable
Marine borer resistance:	Probably resistant *
Treatability of sapwood:	Moderately resistant
Treatability of heartwood:	Resistant

Source:	Handbook of hardwoods (HMSO, 1997)

Tree characteristics, working properties and uses: The timber is difficult to work and has a tendency to distort and has poor abrasion resistance. The species does not have a proven track record in the UK although its technical properties would render it suitable for most outdoor uses where there is a low risk of biodeterioration. The timber is suitable for sea defences on sandy beaches with low abrasion and free from marine borers. The tree is reported to reach a height of 45 m with a diameter of 0.9-1.2 m. The bole is straight and cylindrical.

Availability: Rarely available in the UK but abundant supplies in West Africa.

*Comparable performance to greenheart when challenged with Limnoria spp. under laboratory conditions

Danta	**Combretodendron africanum**		
Other names:	Otutu (Nigeria), Kotibe (Cote d'Ivoire)		
Source of supply:	West Africa		

Available sizes:	Length	Up to 7 600 mm
	Width	Up to 450 mm
	Thickness	Up to 450 mm

Mechanical properties:		*Green*	*Dry*
	Bending strength (N/mm^2)		137
	Modulus of Elasticity (kN/mm^2)		11.7
	Compression strength parallel to grain (N/mm^2)		69.3
	Maximum shear strength parallel to grain (N/mm^2)		21.2
	Impact strength (m)		1.09

Source:	Handbook of hardwoods (HMSO, 1997)
Density:	kg/m^3
Movement:	Medium

Natural durability:	Durable
Marine borer resistance:	Probably resistant
Treatability of sapwood:	Moderately resistant
Treatability of heartwood:	Resistant

Source:

Tree characteristics, working properties and uses:	A heavy timber having good strength properties. Relatively easy to work. The timber has good resistance to abrasion. The tree is reported to reach a height of 27-30 m with a diameter of 0.6-0.8 m. The bole is reasonably straight and cylindrical, above short buttresses.
Availability:	Supplies readily available

Denya   **Cylicodiscus gabunensis**

Other names: Okan

Source of supply: West Africa

Available sizes: Length   Up to 9 000 mm
                                   Width   150-350 mm
                                   Thickness   50-250 mm

Mechanical properties: *Green*   *Dry*
                                   Bending strength (N/mm^2)   101   140
                                   Modulus of Elasticity (kN/mm^2) 12.8   16.1
                                   Compression strength parallel   56.7   85.4
                                   to grain (N/mm^2)
                                   Maximum shear strength
                                   parallel to grain (N/mm^2)
                                   Impact strength (m)

Source:

Density: kg/m^3 960

Movement: No information

Natural durability: Very durable
Marine borer resistance: Resistant *
Treatability of sapwood: Resistant
Treatability of heartwood: Extremely resistant

Source: Handbook of hardwoods (HMSO, 1997)

Tree characteristics, working   A large tree with an average height of 55-60 m, with a
properties and uses: diameter up to 2.5-3.0 m. Difficult to obtain a clean
                                   finish due to interlocked grain. A heavy timber used
                                   for piling and wharf decking, sea defences.

Availability: Available by forward shipment. Good stocks available
                                   in Africa.

*Performed well in laboratory screening trials against Limnoria spp.

Determa	**Ocotea rubra**		
Other names:	Red louro		
Source of supply:	Guyana, Surinam, French Guiana and Brazil		
Available sizes:	Length Width Thickness		
Mechanical properties:		*Green*	*Dry*
	Bending strength (N/mm^2) Modulus of Elasticity (kN/mm^2) Compression strength parallel to grain (N/mm^2) Maximum shear strength parallel to grain (N/mm^2) Impact strength (m)		
Source:			
Density:	kg/m^3		640
Movement:			
Natural durability:	Durable		
Marine borer resistance:	Probably not resistant		
Treatability of sapwood:	Probably resistant		
Treatability of heartwood:	Probably resistant		
Source:	Timbers of the world (TRADA)		
Tree characteristics, working properties and uses:	The tree grows to a height of 30 m with a diameter ranging from 0.6 m to 1.0 m. The boles are cylindrical and clear but may taper considerably. The species is not to be confused with greenheart although they are botanically similar. In terms of strength, the timber is similar to oak. The timber has been used in marine constructions in South America but in areas where there is low risk of Teredo spp.		
Availability:	Available		

Douglas fir	**Pseudotsuga menziesii**		
Other names:	Oregon pine, Columbian pine, British Columbia pine		
Source of supply:	Canada, USA and UK		

Available sizes:	Length	Up to 12 000 mm
	Width	150-350 mm
	Thickness	75-350 mm

Mechanical properties:		*Green*	*Dry*
	Bending strength (N/mm^2)	54	93
	Modulus of Elasticity (kN/mm^2)	10.4	12.7
	Compression strength parallel to grain (N/mm^2)	25.9	52.1
	Maximum shear strength parallel to grain (N/mm^2)	7.2	10.8
	Impact strength (m)	0.66	0.86

Source:	The strength properties of timber (HMSO, 1983)		

Density:	kg/m^3	673	545

Movement:	Small		

Natural durability:	Moderately durable
Marine borer resistance:	No resistance
Treatability of sapwood:	Moderately resistant
Treatability of heartwood:	Resistant

Source:	Handbook of softwoods (HMSO, 1972)

Tree characteristics, working properties and uses: The timber has been used widely in heavy constructions and is suitable for marine and fresh water structures although it must be incised and preservative-treated. British grown Douglas fir is believed to be less durable and slightly weaker. Douglas fir is easy to work and is easy to dry and has little tendency to distort and split. The timber has low resistance to abrasion.

Availability: Readily available

Ekki	**Lophira alata**	
Other names:	Azobe, Kaku, Bongossi	
Source of supply:	Widespread throughout West and Central Africa	

Available sizes:	Length	Up to 8 000 mm
	Width	Up to 100-350 mm
	Thickness	Up to 75-350 mm

Mechanical properties:		*Green*	*Dry*
	Bending strength (N/mm^2)	118	164
	Modulus of Elasticity (kN/mm^2)	13.6	16.2
	Compression strength parallel to grain (N/mm^2)	65.6	85
	Maximum shear strength parallel to grain (N/mm^2)	15.2	23.5
	Impact strength (m)	1.35	1.2

Source:	The strength properties of timber (HMSO, 1983)		
Density:	kg/m^3	1292	1051
Movement:	Large		

Natural durability:	Very durable
Marine borer resistance:	Resistant
Treatability of sapwood:	Moderately resistant
Treatability of heartwood:	Extremely resistant

Source:	Handbook of hardwoods (HMSO, 1997)

Tree characteristics, working properties and uses:	The timber has long been recognised for its excellent properties for heavy constructions and is used widely throughout Europe in marine structures. The timber has excellent resistance to abrasion but is difficult to work and has a tendency to distort and split. The timber has a transition zone between the heartwood and sapwood. This is usually slightly paler than the heartwood. The designer should be aware that this transition wood is only moderately durable. The tree is reported to reach a height of 55 m with a diameter of 1.5-1.8 m. The bole is long and clear.
Availability:	Readily available

Essia   **Combretodendron africanum**

Other names:	Owewe (Nigeria), Minzu (Congo-Brazzaville), Abale (Cote d'Ivoire), Abine (Gabon)
Source of supply:	Widespread throughout West Africa
Available sizes:	Length Width Thickness

Mechanical properties:

	Green	Dry
Bending strength (N/mm^2)		140
Modulus of Elasticity (kN/mm^2)		14.5
Compression strength parallel to grain (N/mm^2)		74.9
Maximum shear strength parallel to grain (N/mm^2)		15.4
Impact strength (m)		1.14

Source:	The strength properties of timber (HMSO, 1983)	
Density:	kg/m^3	960 865
Movement:	Large	

Natural durability:	Probably moderately durable
Marine borer resistance:	Not resistant *
Treatability of sapwood:	Extremely resistant
Treatability of heartwood:	Extremely resistant
Source:	Handbook of hardwoods (HMSO, 1997)
Tree characteristics, working properties and uses:	Prone to degrade when drying. Suitable for rough, heavy construction work. Difficult to work and prone to splitting. The tree is reported to reach a height of 37 m or more with a diameter of 0.8-1.1 m. The bole is reasonably straight and cylindrical.
Availability:	No reliable records of its use in the UK

*Poor performance against Limnoria spp. during laboratory trials

Favinha	**Enterolobium schoenburgkii**		
Other names:	Batibatra		
Source of supply:	Brazil, Guyana		
Available sizes:	Length		
	Width		
	Thickness		

Mechanical properties:		*Green*	*Dry*
	Bending strength (N/mm^2)		128
	Modulus of Elasticity (kN/mm^2)		14.6
	Compression strength parallel to grain (N/mm^2)		66
	Maximum shear strength parallel to grain (N/mm^2)		
	Impact strength (m)		

Source:	Ecotimber handbook		
Density:	kg/m^3	1100	830
Movement:			
Natural durability:	Very durable		
Marine borer resistance:	Probably not resistant *		
Treatability of sapwood:	Probably resistant		
Treatability of heartwood:	Probably resistant		
Source:			

Tree characteristics, working properties and uses:	Timber is very heavy but works fairly well, requires care with finishing. It is used for flooring, decking, civil construction, carpentry and joinery.
Availability:	Regular supply

*Performed poorly against Limnoria spp. during laboratory screening trials

Freijo	**Cordia goeldiana**

Other names: Jenny wood (USA), Cordia wood, Frei jorge (Brazil)

Source of supply: Brazil

Available sizes: Length
Width
Thickness

Mechanical properties:

	Green	Dry
Bending strength (N/mm^2)		97
Modulus of Elasticity (kN/mm^2)		12
Compression strength parallel to grain (N/mm^2)		54.3
Maximum shear strength parallel to grain (N/mm^2)		10.2
Impact strength (m)		

Source: The strength properties of timber (HMSO, 1983)

Density:	kg/m^3	700	585

Movement: n/a

Natural durability:	Durable
Marine borer resistance:	Probably resistant
Treatability of sapwood:	Probably resistant
Treatability of heartwood:	Probably resistant

Source: Handbook of hardwoods (HMSO, 1997)

Tree characteristics, working properties and uses: Similar strength properties to Myannmar teak. Easy to work. The tree is reported to reach a height of 30 m with a diameter of 0.6-1.0 m.

Availability: Poor quality of imports has prevented acceptance of the species in the European market. Limited availability.

Greenheart	**Ocotea rodiaei**	
Other names:		
Source of supply:	Guyana	
Available sizes:	Length	up to 24 300 mm
	Width	up to 200 mm-450 mm
	Thickness	up to 50 mm-450 mm

Mechanical properties:		*Green*	*Dry*
	Bending strength (N/mm^2)	144	190
	Modulus of Elasticity (kN/mm^2)	15.9	21.9
	Compression strength parallel to grain (N/mm^2)	72.5	985
	Maximum shear strength parallel to grain (N/mm^2)	15	20.8
	Impact strength (m)	1.3	1.4

Source:	The strength properties of timber (HMSO, 1983)		
Density:	kg/m^3	1250	1009
Movement:	Medium		

Natural durability:	Very durable
Marine borer resistance:	Resistant
Treatability of sapwood:	Resistant
Treatability of heartwood:	Extremely resistant

Source: Handbook of hardwoods (HMSO, 1997)

Tree characteristics, working properties and uses: The timber has long been recognised for its excellent properties for heavy constructions and is used widely throughout the UK in marine structures. The timber has excellent resistance to abrasion but is difficult to work and has a tendency to distort and split. The tree is reported to reach a height of 21-40 m with a diameter of 1.0 m. The bole is long, straight and cylindrical.

Availability: Readily available

Guariuba	**Claricia racemosa**		
Other names:	Turupay, Amarillo		
Source of supply:	Brazil		
Available sizes:	Length Width Thickness		
Mechanical properties:		*Green*	*Dry*
	Bending strength (N/mm^2)		117
	Modulus of Elasticity (kN/mm^2)		13.7
	Compression strength parallel to grain (N/mm^2)		68
	Maximum shear strength parallel to grain (N/mm^2)		
	Impact strength (m)		
Source:	Ecotimber handbook and South American timbers, 1979 CSIRO		
Density:	kg/m^3	>1000	690
Movement:	Low		
Natural durability:	Durable		
Marine borer resistance:	Probably resistant *		
Treatability of sapwood:	Probably resistant		
Treatability of heartwood:	Probably resistant		
Source:	Ecotimber handbook and South American timbers, 1979 CSIRO		
Tree characteristics, working properties and uses:	Timber is heavy, moderate working due to silica. It is used for furniture, exterior and interior joinery, carpentry, flooring and cladding.		
Availability:	Regular, but limited volumes		

*Comparable performance to greenheart when tested against Limnoria spp. during laboratory trials

Heritiera	**Heritiera littoralis**
Other names:	Tulip mangrove
Source of supply:	Australia, Papua New Guinea, Malaysia
Available sizes:	Length Width Thickness

Mechanical properties:		Green	Dry
	Bending strength (N/mm^2)	73-86	114-134
	Modulus of Elasticity (kN/mm^2)		14.2-16.3
	Compression strength parallel to grain (N/mm^2)		62-71
	Maximum shear strength parallel to grain (N/mm^2)		15-16.7
	Impact strength (m)		

Source:	Characteristics, properties and uses of timber. Vol 1. South-East Asia, Northern Australia and the Pacific. CSIRO

Density:	kg/m^3	725-900

Movement:

Natural durability:	Moderately durable
Marine borer resistance:	Listed as resistant to marine borers *
Treatability of sapwood:	Permeable
Treatability of heartwood:	Resistant

Source:	Characteristics, properties and uses of timber. Vol 1. South-East Asia, Northern Australia and the Pacific. CSIRO

Tree characteristics, working properties and uses:	Timber is hard and strong and difficult to work, due to silica. The tree may reach 30 m in height with a bole up to 18 m and 0.8-1.3 m diameter. Used for wharf and bridge building, decking, cladding and joinery.

Availability:	No information on availability in the UK

* Moderate resistance to Limnoria spp. observed during laboratory trials. Presence of silica may offer good resistance to Teredo spp.

Ipê   **Tabebuia spp.**

Other names: Iron wood, Groenhart, Ipê tabaco

Source of supply: Central and South America, Caribbean

Available sizes: Length   Up to 12 000 mm
                            Width   225 mm-350 mm
                            Thickness   100 mm-350 mm

Mechanical properties:		*Green*	*Dry*
Bending strength (N/mm^2)			137
Modulus of Elasticity (kN/mm^2)			13.9
Compression strength parallel to grain (N/mm^2)			77.6
Maximum shear strength parallel to grain (N/mm^2)			19.4
Impact strength (m)			

Source: The strength properties of timber (HMSO, 1983)

Density: kg/m^3 1300   1050

Movement: Medium

Natural durability: Very durable
Marine borer resistance: Probably resistant *
Treatability of sapwood: Permeable
Treatability of heartwood: Extremely resistant

Source: Handbook of hardwoods (HMSO, 1997)

Tree characteristics, working Timber is very heavy, durable and strong, needing
properties and uses: pre-drilling for nailing. It is used for flooring, decking,
                                 heavy and outdoor construction, jetties, sea defences.

Availability: Regular in limited volumes

*Good resistance against Limnoria spp. observed during laboratory trials

Itauba	**Mezilaurus itauba**		
Other names:	Itauba amarela, Itauba abacate, Itauba preta		
Source of supply:	Brazil		
Available sizes:	Length Width Thickness		
Mechanical properties:		*Green*	*Dry*
	Bending strength (N/mm^2)		124
	Modulus of Elasticity (kN/mm^2)		16.2
	Compression strength parallel to grain (N/mm^2)		62
	Maximum shear strength parallel to grain (N/mm^2)		
	Impact strength (m)		
Source:	Ecotimber handbook		
Density:	kg/m^3	1070	750-1000
Movement:	Large		
Natural durability:	Very durable		
Marine borer resistance:	Resistant		
Treatability of sapwood:	Probably resistant		
Treatability of heartwood:	Probably resistant		
Source:	Ecotimber handbook		
Tree characteristics, working properties and uses:	The timber has been used for marine construction in Brazil. The tree reaches a height of 40-50 m with a bole of 10-30 m of diameter 1.5 m. The bole may be cylindrical or buttressed.		
Availability:	No information on availability in the UK		

APPENDIX 1

Jarana   **Lecythis spp.**

Other names:

Source of supply: Brazil

Available sizes: Length
                            Width
                            Thickness

Mechanical properties:

	Green	Dry
Bending strength (N/mm^2)		152
Modulus of Elasticity (kN/mm^2)		15.6
Compression strength parallel to grain (N/mm^2)		82
Maximum shear strength parallel to grain (N/mm^2)		
Impact strength (m)		

Source: Ecotimber handbook

Density: kg/m^3 1200   850-950

Movement: Large

Natural durability: Very durable
Marine borer resistance: Probably not resistant *
Treatability of sapwood: Probably resistant
Treatability of heartwood: Probably resistant

Source: Ecotimber handbook

Tree characteristics, working properties and uses: A small tree reaching heights of 20 m.

Availability: No information on availability in the UK

*Performed poorly in laboratory trials when challenged with Limnoria spp.

269

Jarrah	**Eucalyptus marginata**	
Other names:		
Source of supply:	South Western Australia	
Available sizes:	Length	5 700 mm
	Width	75 mm-300 mm
	Thickness	50 mm-300 mm

Mechanical properties:		Green	Dry
	Bending strength (N/mm^2)	72	118
	Modulus of Elasticity (kN/mm^2)	9.6	12.1
	Compression strength parallel to grain (N/mm^2)	37.2	63.5
	Maximum shear strength parallel to grain (N/mm^2)	10.3	16.7
	Impact strength (m)	n/a	n/a

Source:	The strength properties of timber (HMSO, 1983)	
Density:	kg/m^3	1009	865

Movement:

Natural durability:	Very durable
Marine borer resistance:	Resistant
Treatability of sapwood:	n/a
Treatability of heartwood:	Extremely resistant

Source: Handbook of hardwoods (HMSO, 1997)

Tree characteristics, working properties and uses: Difficult to work and to nail, pre-boring necessary, holds screws well. This timber has a tendency to distort or split and has medium resistance to abrasion. The tree is reported to reach a height of 30-45 m with a diameter of 1.0 m-1.5 m.

Availability: Readily available

Kapur	**Drylobalanops spp.**	
Other names:		
Source of supply:	South East Asia, principally Malaysia and Indonesia	
Available sizes:	Length	3 000 mm-9 000 mm
	Width	50 mm-300 mm
	Thickness	50 mm-300 mm

Mechanical properties:		*Green*	*Dry*
	Bending strength (N/mm^2)	81	117
	Modulus of Elasticity (kN/mm^2)	10.9	13.0
	Compression strength parallel to grain (N/mm^2)	41.2	69.6
	Maximum shear strength parallel to grain (N/mm^2)	8.5	
	Impact strength (m)	0.71	

Source:	Handbook of hardwoods (HMSO, 1997)		
Density:	kg/m^3	865	700
Movement:	Medium		

Natural durability:	Very durable
Marine borer resistance:	Probably resistant
Treatability of sapwood:	Permeable
Treatability of heartwood:	Extremely resistant

Source: Handbook of hardwoods (HMSO, 1997)

Tree characteristics, working properties and uses: A number of Drylobalanops species are grouped under the commercial description of 'kapur'. The timber is coarse but straight grained. In the green state, kapur is slightly superior in terms of strength than myannmar. The tree is reported to reach a height of 60 m or sometimes more with a diameter of 1.0 m-1.5 m.

Availability: No information on availability in the UK. Supplies are plentiful in S.E. Asia. Possible to procure by arrangement with shipper through timber importer.

Karri	**Eucalyptus diversicolor**	
Other names:		
Source of supply:	South Western Australia	
Available sizes:	Length	Up to 8 500 mm
	Width	75 mm-300 mm
	Thickness	50 mm-300 mm

Mechanical properties:		Green	Dry
	Bending strength (N/mm^2)	77	139
	Modulus of Elasticity (kN/mm^2)	13.4	17.9
	Compression strength parallel to grain (N/mm^2)	37.6	74.5
	Maximum shear strength parallel to grain (N/mm^2)	10.4	16.7
	Impact strength (m)	n/a	n/a

Source:	The strength properties of timber (HMSO, 1983)		
Density:	kg/m^3	1041	913

Movement:

Natural durability:	Durable
Marine borer resistance:	Probably not resistant
Treatability of sapwood:	Probably resistant
Treatability of heartwood:	Extremely resistant

Source: Handbook of hardwoods (HMSO, 1997)

Tree characteristics, working properties and uses: Difficult to work and to nail, pre-boring necessary. This timber has a tendency to distort or split and has medium resistance to abrasion. Little reported use as pilings. The tree is reported to reach a height of 45-60 m or sometimes larger with a diameter of 1.8-3.0 m. The bole is clear.

Availability: Readily available

Kauta/ Kautaballi/Marish	**Licania spp.**		

Other names: Marish-marishballi (Guyana), Anaura (Brazil), Kauston (Surinam), Gris-gris (French Guiana), Bois galettes (French Guiana)

Source of supply: Guyana, Surinam, French Guiana, Brazil

Available sizes: Length
Width
Thickness

Mechanical properties:		*Green*	*Dry*
Bending strength (N/mm^2)			
Modulus of Elasticity (kN/mm^2)			
Compression strength parallel to grain (N/mm^2)			
Maximum shear strength parallel to grain (N/mm^2)			
Impact strength (m)			
Density: kg/m^3		1150-1460	930-1280

Movement:

Natural durability: Moderately durable
Marine borer resistance: Resistant
Treatability of sapwood: Resistant
Treatability of heartwood: Resistant

Source: South American timbers 1979, CSIRO

Tree characteristics, working properties and uses: At least nine species comprise the species group Kauta, Kautaballi and Marish. The timbers are all similar in that they are hard, heavy and strong. High silica content. The timber has long been recognised for its resistance to marine borers and in trials in the USA, they outperformed greenheart, manbarklak and basralocus. The timber is thought to be comparable to greenheart for strength, although inferior in terms of impact resistance.

Availability: No information although thought to be plentiful in supply from South America by prior arrangement with importer.

Keledang	**Artocarpus spp.**		
Other names:			
Source of supply:	South East Asia, principally Malaysia and Indonesia		
Available sizes:	Length Width Thickness		
Mechanical properties:		*Green*	*Dry*
	Bending strength (N/mm^2) Modulus of Elasticity (kN/mm^2) Compression strength parallel to grain (N/mm^2) Maximum shear strength parallel to grain (N/mm^2) Impact strength (m)		
Source:	Timbers of the world (TRADA)		
Density:	kg/m^3	680-960	560-800
Movement:			
Natural durability:	Moderately durable		
Marine borer resistance:	Probably resistant		
Treatability of sapwood:	Probably resistant		
Treatability of heartwood:	Probably resistant		
Source:			
Tree characteristics, working properties and uses:	Used for heavy constructions. Strength varies because of variation of species that comprise keledang but is believed to be similar to that of kapur. Timber has deep interlocked grain and a coarse texture.		
Availability:	No information on availability in the UK. Supplies are plentiful in S.E. Asia. Possible to procure by arrangement with shipper through timber importer.		

Kempas	**Koompassia malaccensis**		
Other names:	Mengeris, Tualang		
Source of supply:	Malaysia, Indonseia		
Available sizes:	Length	3 000 mm	
	Width	150 mm	
	Thickness	50 mm	

Mechanical properties:		*Green*	*Dry*
	Bending strength (N/mm^2)	105	128
	Modulus of Elasticity (kN/mm^2)	15.66	18.6
	Compression strength parallel to grain (N/mm^2)	56.7	68.1
	Maximum shear strength parallel to grain (N/mm^2)		
	Impact strength (m)		

Source:	Handbook of hardwoods (HMSO, 1997)		
Density:	kg/m^3	1024-1100	770-1000

Movement:

Natural durability:	Durable
Marine borer resistance:	Resistant
Treatability of sapwood:	Probably resistant
Treatability of heartwood:	Resistant

Source: Handbook of hardwoods (HMSO, 1997)

Tree characteristics, working properties and uses: Works well but has a tendency to blunt cutters and saws. Difficult to nail. Little reported use as pilings. The tree is reported to reach a height of 55 m. The bole is columnar, with large buttresses.

Availability: Limited in the UK

Keranji	**Dialum spp.**
Other names:	Malaysia
Source of supply:	Malaysia
Available sizes:	Length Width Thickness

Mechanical properties:		*Green*	*Dry*
	Bending strength (N/mm^2)		
	Modulus of Elasticity (kN/mm^2)		
	Compression strength parallel to grain (N/mm^2)		
	Maximum shear strength parallel to grain (N/mm^2)		
	Impact strength (m)		

Source:	
Density:	kg/m^3
Movement:	

Natural durability:	Moderately durable
Marine borer resistance:	Probably resistant
Treatability of sapwood:	Probably resistant
Treatability of heartwood:	Probably resistant

Source:	Timbers if the world (TRADA)
Tree characteristics, working properties and uses:	Keranji is a commercial description applied to a range of dialum species with similar properties. Keranji is a tough, hard, dense timber suitable for heavy construction. It is difficult to work.
Availability:	No information on availability in the UK. Supplies are plentiful in S.E. Asia. Possible to procure by arrangement with shipper through timber importer.

Keruing	**Dipterocarpus spp.**

Other names: Gurjun, Yang, Apitong, Eng

Source of supply: Malaysia, Indonesia, Philippines, Vietnam, Myannmar, Laos, Cambodia

Available sizes:

Length	up to 7 600 mm
Width	50 mm-100 mm (sometimes wider)
Thickness	25 mm-100 mm (sometimes thicker)

Mechanical properties:

	Green	Dry
Bending strength (N/mm^2)	70-83	110-144
Modulus of Elasticity (kN/mm^2)	11.8-16.0	13.7-17.6
Compression strength parallel to grain (N/mm^2)	35.4-43.1	59.6-79.8
Maximum shear strength parallel to grain (N/mm^2)	7.2-8.1	12.2-14.3
Impact strength (m)	0.69-0.81	0.97-1.09

Source: The strength properties of timber (HMSO, 1983)

Density: kg/m^3 753-929   641-849

Movement:

Natural durability: Moderately durable
Marine borer resistance: Probably not resistant
Treatability of sapwood: Probably resistant
Treatability of heartwood: Variable – moderately resistant

Source: Handbook of hardwoods (HMSO, 1997)

Tree characteristics, working properties and uses: Variable timber, but works relatively well. Pre-bore prior to nailing. Tendency to distort on drying. Little reported use as pilings. Medium to high resistance to abrasion. The tree is reported to reach a height of 30-60 m with a diameter of 1.0-1.8 m. The bole is reasonably straight and clear.

Availability: Readily available

Kopie	**Goupia glabra**
Other names:	Kabukalli, Cupiuba (Brazil), Goupi (Guyana), Kopie (Surinam)
Source of supply:	South America
Available sizes:	Length Width Thickness

Mechanical properties:		*Green*	*Dry*
	Bending strength (N/mm^2)		122
	Modulus of Elasticity (kN/mm^2)		14.7
	Compression strength parallel to grain (N/mm^2)		62
	Maximum shear strength parallel to grain (N/mm^2)		
	Impact strength (m)		

Source:	Ecotimber handbook
Density:	kg/m^3 832
Movement:	
Natural durability:	Durable
Marine borer resistance:	Probably not resistant
Treatability of sapwood:	Probably resistant
Treatability of heartwood:	Probably resistant
Source:	Ecotimber handbook
Tree characteristics, working properties and uses:	Reported to be suitable for heavy constructions in water free of Teredo. Kopie is a hard strong timber but, for its density, has lower impact resistance than would be expected. The tree reaches a height of 30-40 m with a bole length of 10-30 m and diameter up to 1.5 m. The bole is buttressed.
Availability:	Limited supplies, Sizes and volumes should be confirmed prior to procurement.

Larch	**Larix spp.**	
Other names:	European, Japanese, Dunkeld/Hybrid larch	
Source of supply:	Mostly UK and some Europe	
Available sizes:	Length	Up to 9 000 mm
	Width	Up to 250 mm
	Thickness	Up to 250 mm

Mechanical properties:		*Green*	*Dry*
	Bending strength (N/mm^2)	43-53	77-92
	Modulus of Elasticity (kN/mm^2)	5.9-6.8	8.5-9.9
	Compression strength parallel to grain (N/mm^2)	19.2-24.3	39.1-46.7
	Maximum shear strength parallel to grain (N/mm^2)	5.9-6.9	11.6-12.4
	Impact strength (m)	0.69-0.86	0.64-0.76

Source:	The strength properties of timber (HMSO, 1983)		
Density:	kg/m^3	580-670	460-540

Movement:

Natural durability:	Moderately durable
Marine borer resistance:	Probably not resistant
Treatability of sapwood:	Moderately resistant
Treatability of heartwood:	Resistant

Source: Handbook of softwoods (HMSO, 1972)

Tree characteristics, working properties and uses: Relatively easy to work, may split on nailing. Low resistance to abrasion.

Availability: Readily available

Louro Gamela   **Nectandra rubra**

Other names: Louro vermelho, Red louro

Source of supply: Brazil

Available sizes: Length
                          Width
                          Thickness

Mechanical properties:		Green	Dry
Bending strength (N/mm^2)			90
Modulus of Elasticity (kN/mm^2)			11.4
Compression strength parallel to grain (N/mm^2)			51
Maximum shear strength parallel to grain (N/mm^2)			
Impact strength (m)			

Source: Ecotimber handbook

Density:	kg/m^3	1000	660

Movement:

Natural durability: Durable
Marine borer resistance: Probably resistant *
Treatability of sapwood: Moderately resistant
Treatability of heartwood: Extremely resistant

Source: Timbers of the world (TRADA)

Tree characteristics, working properties and uses: Timber is medium to heavy, but works fairly well. It is used for decking, shipbuilding, interior and exterior joinery, cladding.

Availability: Regular in large volumes

*Good resistance to Limnoria spp. observed during laboratory trials

Louro Itauba	**Mezilaurus itauba**		
Other names:	Louro itauba		
Source of supply:	South America		
Available sizes:	Length Width Thickness		
Mechanical properties:		*Green*	*Dry*
	Bending strength (N/mm^2)		124
	Modulus of Elasticity (kN/mm^2)		16.2
	Compression strength parallel to grain (N/mm^2)		62
	Maximum shear strength parallel to grain (N/mm^2)		
	Impact strength (m)		
Source:	Ecotimber handbook		
Density:	kg/m^3	1070	855
Movement:	Reported to be low		
Natural durability:	Very durable		
Marine borer resistance:	Probably not resistant *		
Treatability of sapwood:	Probably resistant		
Treatability of heartwood:	Extremely resistant		
Source:	Ecotimber handbook		
Tree characteristics, working properties and uses:	Timber is very heavy, but works fairly well. It is used for flooring, decking, heavy and hydraulic construction interior and exterior joinery, cladding.		
Availability:	Regular in moderate volumes		

*Performed poorly in laboratory trials when challenged with Limnoria spp.

Louro Preto	**Ocotea fagantissima**		
Other names:	Silverballi		
Source of supply:	Brazil		
Available sizes:	Length Width Thickness		
Mechanical properties:		*Green*	*Dry*
	Bending strength (N/mm^2)		89
	Modulus of Elasticity (kN/mm^2)		11.0
	Compression strength parallel to grain (N/mm^2)		49
	Maximum shear strength parallel to grain (N/mm^2)		
	Impact strength (m)		
Source:	Ecotimber handbook		
Density:	kg/m^3	900	540
Movement:	Low		
Natural durability:	Moderately durable – durable		
Marine borer resistance:	Little resistance to marine borers *		
Treatability of sapwood:	Probably resistant		
Treatability of heartwood:	Resistant		
Source:	Ecotimber handbook		
Tree characteristics, working properties and uses:	Timber is medium heavy, but works fairly well, machining may be difficult due to interlocked grain. It is used for flooring and decking, interior and exterior joinery, cladding.		
Availability:	Regular		

*Performed poorly in laboratory trials when challenged with Limnoria spp.

Manbarklak	Escweilera spp.		
Other names:	Kakaralli (Guyana)		
Source of supply:	Surinam, Guyana and to a lesser extent, West Indies, Caribbean		
Available sizes:	Length Width Thickness	3600 mm-7200 mm up to 225 mm up to 225 mm	

Mechanical properties:		*Green*	*Dry*
	Bending strength (N/mm^2) Modulus of Elasticity (kN/mm^2) Compression strength parallel to grain (N/mm^2) Maximum shear strength parallel to grain (N/mm^2) Impact strength (m)		
Source:			
Density:	kg/m^3	1200	1050
Movement:	n/a		
Natural durability: *Marine borer resistance:* *Treatability of sapwood:* *Treatability of heartwood:*	Durable Probably resistant Extremely resistant Extremely resistant		
Source:	Timbers of the world (TRADA)		

Tree characteristics, working properties and uses:	About 80 species of Escweilera spp. are to be found in the Caribbean and South America although few contribute to timber marketed as manbarklak. Manbarklak is extremely hard, tough and strong and has good abrasion resistance. It improves in strength considerably during drying and compares favourably with greenheart although it is inferior in terms of stiffness and tension across the grain. Nevertheless, manbarklak has excellent impact resistance.
Availability:	Limited supplies in the UK although good supplies in South America. Sizes and volumes should be confirmed prior to procurement.

| **Mangrove cedars** | **Xylocarpus granatum & X. Australasicum** |

Other names: Cedar mangrove

Source of supply: Papua New Guinea, Australia, Burma, Philippines

Available sizes:
Length
Width
Thickness

Mechanical properties:

	Green	Dry
Bending strength (N/mm^2)	43-62	67-94
Modulus of Elasticity (kN/mm^2)		9.1-12.4
Compression strength parallel to grain (N/mm^2)		40-53
Maximum shear strength parallel to grain (N/mm^2)		10.3-13.1
Impact strength (m)		

Source: Characteristics, properties and uses of timber. Vol 1. South-East Asia, Northern Australia and the Pacific. CSIRO

Density: kg/m^3 575-640

Movement:

Natural durability: Moderately durable
Marine borer resistance: Reported resistant to Teredo spp.*
Treatability of sapwood: Probably resistant
Treatability of heartwood: Probably resistant

Source:

Tree characteristics, working properties and uses: Timber works easily with all tools. The tree may reach 15-20 m in height with a short straight bole up to 10 m and 0.8 m diameter. Used for salt-water piling, oars and sliced veneer.

Availability: No information on availability in the UK

*Similar performance to greenheart when challenged with Limnoria spp. under laboratory conditions

Massaranduba	**Manilkara spp.**		

Other names: Bolletri, Bullet wood, Balata rouge

Source of supply: Brazil, French Guyana, Surinam

Available sizes:

Length	Up to 12 000 mm	
Width	Up to 350 mm	
Thickness	Up to 350 mm	

Mechanical properties:

	Green	*Dry*
Bending strength (N/mm^2)	154	n/a
Modulus of Elasticity (kN/mm^2)	20.6	n/a
Compression strength parallel to grain (N/mm^2)	75.7	n/a
Maximum shear strength parallel to grain (N/mm^2)	17.3	n/a
Impact strength (m)	n/a	n/a

Source: Ecotimber handbook

Density:

kg/m^3	1120-1400	930-1161

Movement:

Natural durability:	Very durable
Marine borer resistance:	Probably resistant
Treatability of sapwood:	Probably resistant
Treatability of heartwood:	Probably resistant

Source: Handbook of hardwoods (HMSO, 1997)

Tree characteristics, working properties and uses: Difficult to work. Screws and nails well with pre-boring. High resistance to abrasion.

Availability: Available

Mecrusse	**Androstachys johnsonii**		
Other names:			
Source of supply:	Mozambique and Zimbabwe		
Available sizes:	Length		
	Width		
	Thickness		
Mechanical properties:		*Green*	*Dry*
	Bending strength (N/mm^2)		
	Modulus of Elasticity (kN/mm^2)		
	Compression strength parallel to grain (N/mm^2)		
	Maximum shear strength parallel to grain (N/mm^2)		
	Impact strength (m)		
Source:			
Density:	kg/m^3	1200	1000
Movement:	Medium		
Natural durability:	Very durable		
Marine borer resistance:	Probably resistant		
Treatability of sapwood:	Extremely resistant		
Treatability of heartwood:	Extremely resistant		
Source:	Timbers of the world (TRADA)		
Tree characteristics, working properties and uses:	A tall straight tree, height up to 36 m, average diameter up to 0.5 m.		
Availability:	No information on availability in the UK		

Mora	**Mora spp.**		
Other names:			
Source of supply:	Surinam, Guyana, French Guiana, Venezuela, Ecuador, Brazil		
Available sizes:	Length Width Thickness		

Mechanical properties:		*Green*	*Dry*
	Bending strength (N/mm^2)	94	168
	Modulus of Elasticity (kN/mm^2)	14.8	19.2
	Compression strength parallel to grain (N/mm^2)	49.4	87.6
	Maximum shear strength parallel to grain (N/mm^2)	12.5	20
	Impact strength (m)	0.86	1.35

Source:	The strength properties of timber (HMSO, 1983)		
Density:	kg/m^3	1137-1216	993
Movement:			

Natural durability:	Durable
Marine borer resistance:	Probably resistant
Treatability of sapwood:	Moderately resistant
Treatability of heartwood:	Extremely resistant

Source:	Timbers of the world (TRADA)

Tree characteristics, working properties and uses:	Difficult to work timber. Pre-bore prior to nailing. Tendency to distort or split. High resistance to abrasion. Little reported use as pilings. The tree reaches a height of 30-45 m. Buttresses up to 4.5 m, clear bole above, usually cylindrical but sometimes flattened. The sapwood band of mora is wide and this can reduce the cross section of naturally durable material.
Availability:	Readily available

Muiracatiara   **Astronium lecontei/graveolens/fraxinifolium**

Other names: Gonçalo Alves, Zebrawood, Tigerwood

Source of supply: Brazil

Available sizes: Length
                                    Width
                                    Thickness

Mechanical properties:		Green	Dry
Bending strength (N/mm^2)			133
Modulus of Elasticity (kN/mm^2)			17.1
Compression strength parallel to grain (N/mm^2)			76
Maximum shear strength parallel to grain (N/mm^2)			
Impact strength (m)			

Source: Ecotimber handbook

Density:	kg/m^3	1100	880

Movement:

Natural durability: Very durable
Marine borer resistance: Probably moderately resistant *
Treatability of sapwood: Probably resistant
Treatability of heartwood: Extremely resistant

Source: Ecotimber handbook

Tree characteristics, working   The timber is very heavy, but works fairly well, it
properties and uses: does require pre-drilling for nailing. It is used for
                                    flooring, decking, civil and marine construction,
                                    interior and exterior joinery.

Availability: Regular

*Similar performance to greenheart when challenged with Limnoria spp. under laboratory conditions

Muninga	**Pterocarpus angolensis**		
Other names:	Mninga (Tanzania), Ambila (Mozambique), Mukwa (Zambia and Zimbabwe), Kiaat, Kajat, Kajahout (S. Africa)		
Source of supply:	South and Central Africa		
Available sizes:	Length Width Thickness		
Mechanical properties:		*Green*	*Dry*
	Bending strength (N/mm^2)	85	94
	Modulus of Elasticity (kN/mm^2)	7.6	8.4
	Compression strength parallel to grain (N/mm^2)	40.6	57.1
	Maximum shear strength parallel to grain (N/mm^2)		
	Impact strength (m)		
Source:			
Density:	kg/m^3	1073	500-780
Movement:	Small		
Natural durability:	Very durable		
Marine borer resistance:	Probably resistant		
Treatability of sapwood:	Extremely resistant		
Treatability of heartwood:	Extremely resistant		
Source:	Handbook of hardwoods (HMSO, 1997)		
Tree characteristics, working properties and uses:	Reported to have excellent properties for marine constructions. The tree is relatively small, therefore long large sections are difficult to procure. The tree grows to a height of 21 m, but usually 15 m, with a bole length of 4-8 m and diameter about 0.6 m.		
Availability:	No information on availability in the UK		

Muhuhu	**Brachylaena lutchinsii**		
Other names:	Muhugwe (Tanzania)		
Source of supply:	Tanzania and Kenya		
Available sizes:	Length Width Thickness		
Mechanical properties:		*Green*	*Dry*
	Bending strength (N/mm^2)	92	112
	Modulus of Elasticity (kN/mm^2)	8.6	10.1
	Compression strength parallel to grain (N/mm^2)	53.6	70.3
	Maximum shear strength parallel to grain (N/mm^2)	17.4	23.4
	Impact strength (m)	0.69	0.56
Source:			
Density:	kg/m^3	1200	880
Movement:	Small		
Natural durability:	Very durable		
Marine borer resistance:	Probably resistant		
Treatability of sapwood:	Extremely resistant		
Treatability of heartwood:	Extremely resistant		
Source:	The strength properties of timber (HMSO, 1983)		
Tree characteristics, working properties and uses:	The tree is 'medium' sized, therefore it is difficult to obtain large section sizes. The timber is hard and dense, difficult to work but has excellent resistance to abrasion. The tree reaches a height of 25 m, but commonly only 9-18 m with a fluted and twisted bole length of 2-6 m, and diameter of 0.5-0.6 m.		
Availability:	Limited		

Oak   **Quercus spp.**

Other names:

Source of supply: UK and Europe

Available sizes: Length   Up to 7 600 mm
                                     Width   Up to 300 mm
                                     Thickness   Up to 300 mm

Mechanical properties: *Green*   *Dry*
                                     Bending strength (N/mm^2)   59   97
                                     Modulus of Elasticity (kN/mm^2)   8.3   10.1
                                     Compression strength parallel   27.6   51.6
                                     to grain (N/mm^2)
                                     Maximum shear strength   9.1   13.7
                                     parallel to grain (N/mm^2)
                                     Impact strength (m)   0.86   0.84

Source: The strength properties of timber (HMSO, 1983)

Density: kg/m^3 670-720

Movement:

Natural durability: Very durable
Marine borer resistance: Not resistant
Treatability of sapwood: Permeable
Treatability of heartwood: Resistant

Source: Handbook of hardwoods (HMSO, 1997)

Tree characteristics, working   Relatively easy to work and screws and nails well –
properties and uses: pre-boring will probably be necessary. Tendency to
                                     distort and especially splitting. The tree reaches a
                                     height of 30 m with a bole length of 12-35 m and
                                     diameter 0.9-1.2 m.

Availability: Readily available

Opepe   **Nauclea diderrichii**

Other names: Kussia, Bilinga, Badi

Source of supply: West Africa

Available sizes:

Length	Up to 9 000 mm
Width	Up to 350 mm
Thickness	Up to 350 mm

Mechanical properties:

	Green	Dry
Bending strength (N/mm^2)	94	120
Modulus of Elasticity (kN/mm^2)	11.9	13.4
Compression strength parallel to grain (N/mm^2)	51.6	71.7
Maximum shear strength parallel to grain (N/mm^2)	13.1	17.7
Impact strength (m)	0.81	0.71

Source: The strength properties of timber (HMSO, 1983)

Density:	kg/m^3	945	753

Movement: Small

Natural durability: Very durable
Marine borer resistance: Resistant
Treatability of sapwood: Permeable
Treatability of heartwood: Moderately resistant

Source: The strength properties of timber (HMSO, 1983)

Tree characteristics, working properties and uses: Moderately easy timber to work but may split when nailed, pre-boring is necessary. Medium to high abrasion resistance. The tree reaches a height of 50 m with a bole length of 25-30 m and diameter up to 1.5 m. The bole is long and cylindrical.

Availability: Readily available

Piquia	**Caryocar villosum**		
Other names:	Pequia, Piquia bravo, Vinagreira		
Source of supply:	Brazil		
Available sizes:	Length Width Thickness		
Mechanical properties:		*Green*	*Dry*
	Bending strength (N/mm^2)		123
	Modulus of Elasticity (kN/mm^2)		14.3
	Compression strength parallel to grain (N/mm^2)		66
	Maximum shear strength parallel to grain (N/mm^2)		
	Impact strength (m)		
Source:			
Density:	kg/m^3	1120	810
Movement:	Large		
Natural durability:	Very durable		
Marine borer resistance:	Probably resistant *		
Treatability of sapwood:	Permeable		
Treatability of heartwood:	Moderately resistant		
Source:	Ecotimber handbook		
Tree characteristics, working properties and uses:	Very hard, strong timber, particularly suitable for marine constructions. The tree is reported to be very large with a buttressed bole that is usually tall and straight.		
Availability:	Regular		

*Performed well in laboratory tests when challenged with Limnoria spp.

Piquia Marfim   **Aspidosperma desmathum**

Other names: Araracanga, Shibadan

Source of supply: Brazil

Available sizes: Length
Width
Thickness

Mechanical properties:		Green	Dry
Bending strength (N/mm^2)			172
Modulus of Elasticity (kN/mm^2)			20.9
Compression strength parallel to grain (N/mm^2)			91
Maximum shear strength parallel to grain (N/mm^2)			
Impact strength (m)			

Source: Ecotimber handbook

Density:	kg/m^3	1000	770

Movement:

Natural durability: Very durable
Marine borer resistance: Probably not resistant *
Treatability of sapwood: Probably resistant
Treatability of heartwood: Probably resistant

Source: Ecotimber handbook

Tree characteristics, working properties and uses: Timber is heavy, but works fairly well, and pre-drilling required. It is used for flooring and decking, civil construction, truck floors and tool handles.

Availability: Regular

*Performed poorly in laboratory tests when challenged with Limnoria spp.

Pitch pine   **Pinus caribaea, P. palustris, P. elliotti**

Other names: American/Caribbean, etc. pitch pine according to origin

Source of supply: Caribbean, South East United States

Available sizes:

Length	Up to 10 600 mm
Width	Up to 500 mm
Thickness	Up to 500 mm

Mechanical properties:

	Green	Dry
Bending strength (N/mm^2)	66-83	107
Modulus of Elasticity (kN/mm^2)	10.4-11.4	12.6
Compression strength parallel to grain (N/mm^2)	33-37.4	56.1
Maximum shear strength parallel to grain (N/mm^2)	9.2-9.9	14.3
Impact strength (m)	0.99-1.04	0.91

Source: The strength properties of timber (HMSO, 1983)

Density: kg/m^3 977   769

Movement:

Natural durability: Moderately durable
Marine borer resistance: Probably not resistant
Treatability of sapwood: Permeable
Treatability of heartwood: Moderately resistant

Source: Handbook of softwoods (HMSO, 1972)

Tree characteristics, working properties and uses: Moderately difficult timber to work due to the presence of resins. Pre-boring may be necessary. Tendency to distort or split.

Availability: Available

Purpleheart	**Peltogyne spp.**		
Other names:	Amaranth		
Source of supply:	Brazil, Surinam, French Guiana, Guyana		
Available sizes:	Length	Up to 9 000 mm	
	Width	Up to 300 mm	
	Thickness	Up to 300 mm	

Mechanical properties:

	Green	Dry
Bending strength (N/mm^2)	105	147
Modulus of Elasticity (kN/mm^2)	14	16.7
Compression strength parallel to grain (N/mm^2)	56.5	78.5
Maximum shear strength parallel to grain (N/mm^2)	14.9	18.9
Impact strength (m)	1.04	1.19

Source: The strength properties of timber (HMSO, 1983)

Density: kg/m^3

Movement:

Natural durability:	Durable
Marine borer resistance:	Moderately durable *
Treatability of sapwood:	Permeable
Treatability of heartwood:	Extremely resistant

Source: Handbook of hardwoods (HMSO, 1997)

Tree characteristics, working properties and uses: Difficult timber to work. Pre-boring may be necessary. Tendency to split, especially when nailed. This timber has high resistance to abrasion. The tree commonly reaches a height of 38-45 m and diameter up to 1.5 m, more usually 0.6-0.9 m. Long lengths have a tendency to be sprung or exhibit bowing characteristics after being sawn. This is attributed to reaction wood as the tree can grow very tall. It is recommended not to use the timber in very long lengths to avoid any problems caused by spring/bow.

Availability: Readily available

*Out-performed greenheart when challenged with Limnoria spp. under laboratory conditions

Pyinkado	**Xylia xylocarpa.**		
Other names:	Pyin, Pran, Pkhay, Irul		
Source of supply:	Myannmar and India		
Available sizes:	Length Width Thickness		
Mechanical properties:		*Green*	*Dry*
	Bending strength (N/mm^2)	113	145
	Modulus of Elasticity (kN/mm^2)	14.6	16.1
	Compression strength parallel to grain (N/mm^2)	57.4	79.5
	Maximum shear strength parallel to grain (N/mm^2)		
	Impact strength (m)		
Source:	Handbook of hardwoods (HMSO, 1997)		
Density:	kg/m^3	1150	980
Movement:	Medium		
Natural durability:	Very durable		
Marine borer resistance:	Resistant		
Treatability of sapwood:	Moderately resistant		
Treatability of heartwood:	Extremely resistant		
Source:	Handbook of hardwoods (HMSO, 1997)		
Tree characteristics, working properties and uses:	The timber is difficult to work. The timber is widely distributed in Myannmar and to a lesser extent in India. The tree reaches a height of 30-37 m and diameter 0.8-1.2 m. The bole is straight and fairly cylindrical.		
Availability:	Not available in the UK unless on forward shipment.		

Red Angelim	**Dinizia excelsa**		
Other names:			
Source of supply:	South America		
Available sizes:	Length Width Thickness		
Mechanical properties:		*Green*	*Dry*
	Bending strength (N/mm^2)		159
	Modulus of Elasticity (kN/mm^2)		19.4
	Compression strength parallel to grain (N/mm^2)		90
	Maximum shear strength parallel to grain (N/mm^2)		
	Impact strength (m)		
Source:	Ecotimber handbook		
Density:	kg/m^3	1500	1080-1200
Movement:	Large		
Natural durability:	Very durable		
Marine borer resistance:	Probably resistant		
Treatability of sapwood:	Probably resistant		
Treatability of heartwood:	Probably resistant		
Source:	Timbers of the world (TRADA)		
Tree characteristics, working properties and uses:	The timber is difficult to work. The tree grows to a height of 40-70 m and has a large bole up to 40 m in height with a diameter of 2.5 m. The bole is usually cylindrical.		
Availability:	Readily available		

Red Peroba   **Aspidosperma polyneuron**

Other names:

Source of supply: Brazil

Available sizes: Length
Width
Thickness

Mechanical properties:	Green	Dry
Bending strength (N/mm^2)		88
Modulus of Elasticity (kN/mm^2)		8.3
Compression strength parallel to grain (N/mm^2)		39.8
Maximum shear strength parallel to grain (N/mm^2)		
Impact strength (m)		

Source: The strength properties of timber (HMSO, 1983)

Density: kg/m^3 720-830

Movement: Large

Natural durability: Durable
Marine borer resistance: Probably resistant
Treatability of sapwood: Probably resistant
Treatability of heartwood: Probably resistant

Source: Ecotimber handbook

Tree characteristics, working properties and uses: The timber is difficult to work. The tree is variable in height and girth but on average reaches 27 m, with a diameter of 0.75 m. The bole is usually cylindrical.

Availability: Readily available

Redwood (European)	**Pinus sylvestris**	
Other names:	Scots pine, Red pine, Red deal, Yellow deal, Norway fir, Scots fir, Finnish/Baltic/Siberian, etc. Redwood according to origin.	
Source of supply:	Europe	

Available sizes:	Length	up to 6 000 mm, possibly 7 200
	Width	variable up to 275 mm
	Thickness	variable up to 250 mm

Mechanical properties:		*Green*	*Dry*
	Bending strength (N/mm^2)	44	83
	Modulus of Elasticity (kN/mm^2)	7.7	10
	Compression strength parallel to grain (N/mm^2)	21	45
	Maximum shear strength parallel to grain (N/mm^2)	5.9	11.3
	Impact strength (m)	0.69	0.64-0.76

Source:	The strength properties of timber (HMSO, 1983)		
Density:	kg/m^3	580-670	460-540

Movement:

Natural durability:	Moderately durable
Marine borer resistance:	Not resistant
Treatability of sapwood:	Permeable
Treatability of heartwood:	Moderately resistant

Source:	Handbook of softwoods (HMSO, 1972)

Tree characteristics, working properties and uses:	Relatively easy to work apart from machining knots and if large quantities of resin are present. Slight tendency to distort or split. Little reported use as pilings. Low resistance to abrasion.
Availability:	Readily available

Resak	**Cotylelobium spp. and Viaticum spp.**		
Other names:			
Source of supply:	Common throughout South East Asia		
Available sizes:	Length Width Thickness		
Mechanical properties:		*Green*	*Dry*
	Bending strength (N/mm^2) Modulus of Elasticity (kN/mm^2) Compression strength parallel to grain (N/mm^2) Maximum shear strength parallel to grain (N/mm^2) Impact strength (m)		
Source:			
Density:	kg/m^3	850-1200	700-1000
Movement:	Medium		
Natural durability:	Very durable		
Marine borer resistance:	Resistant		
Treatability of sapwood:	Moderately resistant		
Treatability of heartwood:	Extremely resistant		
Source:	Handbook of hardwoods (HMSO, 1997)		

Tree characteristics, working properties and uses: Resak is a commercial term used to describe a number of species with similar properties. Used for heavy construction and suitable for marine purposes. Durability, strength and abrasion resistance are linked to density. The heavier varieties of resak are superior. The timber is difficult to work.

Availability: The timber is widely distributed in South East Asia. Not available in the UK, may be procurable by forward shipment.

Robinia	**Robinia pseudoacacia**		
Other names:	False acacia, Black locust		
Source of supply:	North America, Europe, Asia		
Available sizes:	Length Width Thickness		
Mechanical properties:		*Green*	*Dry*
	Bending strength (N/mm^2)		133
	Modulus of Elasticity (kN/mm^2)		14.2
	Compression strength parallel to grain (N/mm^2)		71
	Maximum shear strength parallel to grain (N/mm^2)		
	Impact strength (m)		
Source:	Handbook of hardwoods (HMSO, 1997)		
Density:	kg/m^3	900- 1200	720 average
Movement:	High		
Natural durability:	Durable		
Marine borer resistance:	Probably moderately durable *		
Treatability of sapwood:	Probably resistant		
Treatability of heartwood:	Extremely resistant		
Source:	Handbook of hardwoods (HMSO, 1997)		
Tree characteristics, working properties and uses:	Grows to around 24-27 m, with a diameter up to 0.6 m which is often fluted. A hard heavy timber used for posts, stakes, gates, boat planking and weatherboarding. Nailing can be difficult and machining and sawing is satisfactory.		
Availability:	Readily available		

*Comparable performance to greenheart observed when challenged with Limnoria spp. under laboratory conditions

Southern Blue Gum	**Eucalyptus globulus**		
Other names:	Blue gum, Tasmanian gum, Eurabbi		
Source of supply:	Australia		
Available sizes:	Length Width Thickness		
Mechanical properties:		*Green*	*Dry*
	Bending strength (N/mm^2) Modulus of Elasticity (kN/mm^2) Compression strength parallel to grain (N/mm^2) Maximum shear strength parallel to grain (N/mm^2) Impact strength (m)		
Source:			
Density:	kg/m^3		
Movement:	Medium		
Natural durability:	Very durable		
Marine borer resistance:	Resistant		
Treatability of sapwood:	n/a		
Treatability of heartwood:	Resistant		
Source:	Timbers of the world (TRADA)		
Tree characteristics, working properties and uses:	Available in large section sizes and has a good track record for use in the marine environment. Strength properties are similar to karri.		
Availability:	No records of availability in the UK		

Suradan	**Hyeronima spp.**		
Other names:	Suradanni, Pilon (Guyana), Sorodon anoniwana (Surinam), Zapatero (Panama), Nancito (Nicaragua), Margoncalo and Urucurana (Brazil)		
Source of supply:	Central and South America		
Available sizes:	Length Width Thickness		
Mechanical properties:		*Green*	*Dry*
	Bending strength (N/mm^2) Modulus of Elasticity (kN/mm^2) Compression strength parallel to grain (N/mm^2) Maximum shear strength parallel to grain (N/mm^2) Impact strength (m)		
Source:			
Density:	kg/m^3	1185	800
Movement:	Medium		
Natural durability: *Marine borer resistance:* *Treatability of sapwood:* *Treatability of heartwood:*	Moderately durable – very durable Resistant Probably resistant Probably resistant		
Source:	Timbers of the word (TRADA)		
Tree characteristics, working properties and uses:	Collectively, about 25 species comprise the commercial group of suradan. The timber is difficult to work but is suitable for marine construction.		
Availability:	No records of availability in the UK		

Tatajuba	**Bagassa guianensis**		
Other names:	Bagasse (Guyana), Gele bagasse (Surinam)		
Source of supply:	Guyana, French Guiana, Brazil and Surinam		
Available sizes:	Length Width Thickness		
Mechanical properties:		*Green*	*Dry*
	Bending strength (N/mm^2)		121
	Modulus of Elasticity (kN/mm^2)		17.3
	Compression strength parallel to grain (N/mm^2)		78
	Maximum shear strength parallel to grain (N/mm^2)		
	Impact strength (m)		
Source:	The strength properties of timber (HMSO, 1983)		
Density:	kg/m^3	1100	720-840
Movement:	Medium		
Natural durability:	Very durable		
Marine borer resistance:	Moderately resistant		
Treatability of sapwood:	Probably resistant		
Treatability of heartwood:	Probably resistant		
Source:	Timbers of the world (TRADA)		
Tree characteristics, working properties and uses:	The tree reaches a height of 27-30 m and a diameter of 0.6 m. The bole is cylindrical and 18-21 m high. The timber has been used for heavy construction, both civil and marine.		
Availability:	No information on availability in the UK		

Teak	Tectona grandis		
Other names:	Sagwan, Teku, Teka, Kyun		
Source of supply:	Myannmar, India, Thailand and has been introduced as a plantation species in Malaysia, Indonesia, Philippines and Central America. The properties listed below only refer to native teak from natural forests.		
Available sizes:	Length Width Thickness		
Mechanical properties:		*Green*	*Dry*
	Bending strength (N/mm^2)	84	106
	Modulus of Elasticity (kN/mm^2)	8.8	10.0
	Compression strength parallel to grain (N/mm^2)	42.8	60.4
	Maximum shear strength parallel to grain (N/mm^2)		
	Impact strength (m)		
Source:	Handbook of hardwoods (HMSO, 1997) Figures above are for Myannmar teak		
Density:	kg/m^3	817	610-690
Movement:	Small		
Natural durability:	Very durable		
Marine borer resistance:	Moderately resistant		
Treatability of sapwood:	Probably resistant		
Treatability of heartwood:	Extremely resistant		
Source:	Handbook of hardwoods (HMSO, 1997)		
Tree characteristics, working properties and uses:	On favourable sites, the tree can reach a height of 39-45 m and a diameter of 1.5 m with a clear bole of 10-24 m. The timber has been used for a wide range of purposes ranging from shipbuilding and heavy construction.		
Availability:	Good		

Turpentine   **Syncarpia glomulifera**

Other names: Red lustre, Lustre

Source of supply: Queensland and New South Wales

Available sizes: Length
                        Width
                        Thickness

Mechanical properties: *Green*   *Dry*
                        Bending strength (N/mm^2)
                        Modulus of Elasticity (kN/mm^2)
                        Compression strength parallel
                        to grain (N/mm^2)
                        Maximum shear strength
                        parallel to grain (N/mm^2)
                        Impact strength (m)

Source:

Density: kg/m^3 817   610-690

Movement: Small

Natural durability: Very durable
Marine borer resistance: Resistant
Treatability of sapwood: Probably resistant
Treatability of heartwood: Probably resistant

Source: Handbook of hardwoods (HMSO, 1997)

Tree characteristics, working Turpentine is a hard, dense timber, ideal for use in the
properties and uses: marine environment.

Availability: No records of availability in the UK

Uchi	**Sacoglottis guiansis**		
Other names:	Bagasse (Guyana), Gele bagasse (Surinam)		
Source of supply:	Guyana, French Guiana, Brazil and Surinam		
Available sizes:	Length Width Thickness		
Mechanical properties:		*Green*	*Dry*
	Bending strength (N/mm^2)		
	Modulus of Elasticity (kN/mm^2)		20.5
	Compression strength parallel to grain (N/mm^2)		
	Maximum shear strength parallel to grain (N/mm^2)		
	Impact strength (m)		
Source:	Ecotimber handbook		
Density:	kg/m^3	1100-1200	960
Movement:	Large		
Natural durability:	Moderately durable		
Marine borer resistance:	Probably not resistant		
Treatability of sapwood:	Probably resistant		
Treatability of heartwood:	Probably resistant		
Source:	Ecotimber handbook		
Tree characteristics, working properties and uses:	Little information available		
Availability:	Available		

Utile	**Entandrophragma utile**

Other names: Sipo (Cote d'Ivoire), Assie (Cameroon)

Source of supply: West and Central Africa

Available sizes: Length
Width
Thickness

Mechanical properties:

	Green	Dry
Bending strength (N/mm^2)	79	103
Modulus of Elasticity (kN/mm^2)	9 600	10 800
Compression strength parallel to grain (N/mm^2)	38.2	60.4
Maximum shear strength parallel to grain (N/mm^2)	10.8	16.9
Impact strength (m)	0.74	0.74

Source: The strength properties of timber (HMSO, 1983)

Density: kg/m^3 660

Movement: Medium

Natural durability: Moderately durable
Marine borer resistance: Probably not resistant
Treatability of sapwood: Probably resistant
Treatability of heartwood: Probably resistant

Source: Handbook of hardwoods (HMSO, 1997)

Tree characteristics, working properties and uses: Excellent general-purpose construction timber. Strength properties are similar to those of American mahogany. The tree reaches a height of 45 m and a diameter of 2.0 m. The bole is straight, free from buttresses and can reach a height of 24 m.

Availability: Readily available

Wacapou	**Soucapoua americanan**		
Other names:	Bois angelim (French Guiana), Bruinhart of Wacapou (Surinam)		
Source of supply:	South America		
Available sizes:	Length Width Thickness		
Mechanical properties:		*Green*	*Dry*
	Bending strength (N/mm^2) Modulus of Elasticity (kN/mm^2) Compression strength parallel to grain (N/mm^2) Maximum shear strength parallel to grain (N/mm^2) Impact strength (m)		
Source:			
Density:	kg/m^3	1200	945
Movement:			
Natural durability:	Very durable		
Marine borer resistance:	Probably resistant		
Treatability of sapwood:	Probably resistant		
Treatability of heartwood:	Probably resistant		
Source:	Ecotimber handbook		
Tree characteristics, working properties and uses:	The tree is tall, slender and not buttressed. The timber has excellent strength properties and is superior to oak. It is relatively difficult to work.		
Availability:	No records of availability in the UK		

Wallaba   **Eperua falcata, Eperua grandifola**

Other names: Waapa, Bijlhout, Apa, Ituri

Source of supply: Guyana, Surinam, Brazil

Available sizes: Length
Width
Thickness

Mechanical properties:

	Green	*Dry*
Bending strength (N/mm^2)	104	139
Modulus of Elasticity (kN/mm^2)	15.0	14.7
Compression strength parallel to grain (N/mm^2)	57.8	77.3
Maximum shear strength parallel to grain (N/mm^2)		
Impact strength (m)		

Source: Handbook of hardwoods (HMSO, 1997)

Density: kg/m^3 910

Movement:

Natural durability: Very durable
Marine borer resistance: Not resistant
Treatability of sapwood: Probably resistant
Treatability of heartwood: Extremely resistant

Source: Timbers of the world (TRADA)

Tree characteristics, working properties and uses: The tree grows to about 30 m with a diameter of 0.6 m. The timber is very hard and heavy, and gum exudation can be a problem when sawing or on finished surfaces. It is used for wharf and bridge construction, decking and flooring.

Availability: Available

Zebrawood

Astronium graveolens

Other names:	Palo de cera, Palo de culebra, Gusanero, Gateodo, Guarita
Source of supply:	Brazil, Colombia, Venezuela, Central America
Available sizes:	Length Width Thickness

Mechanical properties:

	Green	Dry
Bending strength (N/mm^2)		133
Modulus of Elasticity (kN/mm^2)		17.1
Compression strength parallel to grain (N/mm^2)		76
Maximum shear strength parallel to grain (N/mm^2)		
Impact strength (m)		

Source:	Ecotimber handbook		
Density:	kg/m^3	1100	880
Movement:	Large		

Natural durability:	Very durable
Marine borer resistance:	Probably resistant
Treatability of sapwood:	Probably resistant
Treatability of heartwood:	Probably resistant
Source:	Timbers of the world (TRADA)
Tree characteristics, working properties and uses:	The tree can reach a height of up to 50 m with a bole of 20-30 m and diameter of 1.5 m. Bole is generally clear and cylindrical.
Availability:	No information on availability in the UK

Appendix 2.
Model specification

TIMBER STRUCTURES – A FRAMEWORK FOR A CONSTRUCTION SPECIFICATION

2.1 Preamble

The aim of this specification is to provide a framework for the construction specification of any timber structure used in coastal, harbour or river engineering by the use of a simple iterative process aided by typical and site-specific drawings, and to provide in electronic form a template for the first draft of a Project Specification. The template in its complete and detailed form does not necessarily apply to every situation, but it does provide a reasonably comprehensive list of the details that need to be addressed. To this end, the language used is as it would be in the finished Specification. Designers wishing to use this document as an *aide memoir* may do so by substituting the word 'should' for 'shall' wherever it occurs. For convenience the Specification contains reference to outline details given elsewhere in this document for some frequently used elements.

Information contained in either standard or contract drawings should be referred to in the Specification. References should be in the form of "as Drawing Number … Detail(s)…" not by general reference. Words or phrases in [ ] are suggestions for likely options, ( ) where contract specific detail needs inserting, …where references or numbers need inserting.

Final checks before issue of documents

Cross check that the references to drawings in the Specification agree with the numbering in the final version Issue Drawings.

Check that the names used for structures and members agree with those used in the Bills of Quantity and the Issue Drawings.

The Specification follows a typical construction sequence, starting with the material supply and the Site Compound, and ending in Site Specials. After Section 2.2, each type of main structure should have a separate heading, and the members of each structure should be dealt with in the overall order of sections, e.g. groynes should contain specifications for piles, walings and planking. Raking supports to groynes should be a subsection to groynes: plant bays, end details and tie-ins should be included under Site Specials.

Construction Constraints. These should be summarised within the Construction Specification. The Constraints fall into two categories, General, which affect all or most operations, and Specific, which affect one structure or class of structures. The former should appear as a preamble, the latter in the section(s) they affect. There should be no need to reiterate the constraints in full – just a subject heading and a reference is adequate (i.e. Routing constraints as Dwg ..., Noise constraints as Environmental Statement Clause ..., Working hours as Spec. Clause...) – unless they are only referred to generally elsewhere and full details are laid out in this section. The latter is most likely to apply to Specific Constraints. The aims of summarising all the Constraints in one part of the document are to:

- Give the design team a chance to check the following:
  i)   *All necessary constraints have been included*
  ii) *Constraints are not contradictory*
  iii) *Superfluous constraints have not been included.*
  Any of the above can have an adverse affect on costs, by decreasing buildability, or requiring post tender changes to construction methods.
- Give the tendering contractors' teams a ready point of reference to start planning their methods of working, resourcing them, and preparing submission documents.
- Provide an early reference point for site staff to help the contract to proceed efficiently.

Meanings:
a) Engineer shall mean the person appointed to oversee the Contract on the Client's behalf.
b) "....mm x ....mm" indicates the section dimensions, and may be substituted by "Nominal ....mm x ....mm", "[nominal] ....mm diameter [tapered (minimum dimensions or diameters)]".
c) "....mm - ....mm long" is the length dimension. If all the members being described are the same length, then "....mm long" is appropriate.
d) References to Figures and Sections are to Figures and Sections incorporated in the Manual unless otherwise indicated.

Abbreviations:
Std. Standard
Dwg. Drawing
No. Number

2.2   *Materials*

Materials for works may be procured directly by the Client or included within the construction contract. Whilst the requirements for different types of structures will vary greatly, the following clauses will be applicable to most structures (although not recycled components).

General

All timber shall be of appropriate quality for use in marine and civil engineering construction with ends cut square and appropriately protected. All timber shall be free from [sapwood] [shakes] [knots] [unsound or rotten heart] [holes or plugged defects] (other defects).

Timber shall, where possible, comply with appropriate British Standards (e.g. BS 5756: 1997 *Visual strength grading of hardwood*) and be marked accordingly. Where applicable timber shall comply with [a recognised standard appropriate for the species and size of timber supplied] (e.g. Guyana Grading Rules for Sawn and Baulk Timbers). Copies of such standards shall be supplied to the Engineer.

Physical requirements

It is essential that the required properties are clearly specified, in some situations this may be through the specification of a particular species of timber from a particular source. However, it is preferable to specify the generic properties required for each particular component and allow potential suppliers to provide costs for a range of species fulfilling the requirements. In order to satisfy requirements for responsible procurement it will be necessary for the source and type of timber to be approved by the Engineer before procurement proceeds.

It is suggested that physical requirements for each component are briefly described and that the minimum requirements for each component are summarised in a table of the form shown below. This could be combined with Table 5.3 to provide a comprehensive summary of timber properties at the time of tendering.

Component	Strength class	Minimum density	Abrasion	Resistance to marine borers	Decay
1.					
2.					
3.					
4.					

Dimensional tolerance

Timber shall comply with BS EN 1313 –2. Permitted deviations and preferred sizes except where specified below:

Primary members

Secondary members

Tertiary members
(Different tolerances are likely to be needed for hewn, sawn, thicknessed and planed timbers in various size ranges)

Metal fixings and plates

All steel bolts, nuts, washers, coach screws, plates and fittings [excluding driving rings] not made from stainless steel shall be hot-dipped galvanised in accordance with BS EN ISO 1461.

2.3. Site Compound

a) Part of the Site Compound shall be designated as a Timber Compound for storage and preparation of timber [as shown on layout Dwg. No….]. It shall be clearly demarked from the rest of the Compound.

b) All incoming timber shall be inspected for transit damage and conformity to the Materials Specification before acceptance.

c) Timber shall be handled in such a way as to prevent bruising, warping or cracking.

d) Timber shall be stacked by type and section on bearers placed to prevent distortion during storage, and steps taken to protect it from weather where necessary. Stacks should be limited in height bearing in mind current Health and Safety legislation, risks to operatives and trespassers (especially juveniles), fire risk (especially with small section softwood) and possible warping or other damage to the timber.

e) Everyone working with timber shall have suitable personal protection (see Health & Safety legislation). Where there is a legal requirement for machinery operators to be certified, regular checks shall be made to ensure that all relevant certificates are held and are up to date.

f) For ease of operation and convenience the following preparatory works normally take place in the Compound:

  i) Cutting to length where appropriate.

  ii) Preparation of piles [including fitting shoes and rings as shown in Figure 7.11].

  iii) Where practicable, drilling bolt and dowel holes. Bolt holes shall generally be larger than the nominal bolt diameter [see Specification clause/Drawing Number] [recessed for bolt heads, nuts and washers]. Bolts shall be tightened so that members fit closely and retightened when timbers reach stable moisture content. Dowel holes shall be drilled not greater than the nominal dowel diameter. Bolt and dowel holes shall be drilled complete from one side. See Section 6.6.4

  iv) Where practicable, drill pilot holes for screws, spikes and dogs. Holes for coach screws shall be drilled as [Figure 6.15] [and recessed for heads and washers]. See Section 6.6.4. Smaller screws (such as deck batten fastenings) and square fittings shall be predrilled as [Dwg….].

g) Cut ends of all treated timber to be coated with a suitable (specify) site treatment.

h) Prepared timber shall be stacked separately to unprepared timber.

i) Where possible all offcuts shall be used in the Contract. At the end of the Contract the Engineer shall inspect the remaining offcuts, selecting any suitable for salvage to be disposed of in accordance with the Contract [Clause number…] (or detail arrangements). The balance shall remain the property of the Contractor. [Burning on site will not be permitted.]

j) Where cutting and/or drilling is required in places where there is a risk of equipment coming into contact with water or wet ground, such as the intertidal zone, over water or water margins, it is good and safe practice to limit equipment to non electric tools such as hand, hydraulic or air powered tools.

2.4. Demolition

a) Demolition shall be carried out in such a way as to maximise materials for re-cycling unless otherwise ordered by the Engineer. [In particular ....]

b) Salvaged timber shall be taken to the Timber Compound for inspection, machining and approval for reuse.

c) Reusable timber shall be stacked as for new in stockpiles separate from new timber.

d) At the end of the Contract the remaining reusable timber shall be disposed of [in accordance with Clause...] (or detail arrangements).

e) Where there are problems carrying out demolition as in a), the Contractor shall consult with the Engineer and the Engineer shall instruct the Contractor as to steps to be taken [see Clause...]. {i.e. piles which won't draw to be cut off at substrate/bed level, timbers buried in a disturbance sensitive substrate to be left in situ.}

f) Demolition Constraints.

2.5. Primary members

Member categories are as described in Section 6.6.1 and illustrated in Figure 6.9. Primary members include those contributing to the main stability of the structure (such as piles, posts, ramp/stair supports and the like but generally not fence, railing, notice board or sign board posts, etc.)

a) Piles
    i) Piles for (name of member) shall be ....mm x ....mm (type of timber) ....mm - ....mm long and driven to the depths/set/cap levels shown on Dwg(s). No....
    ii) Piling shall be carried out generally in accordance with Specification for Piling Works and Embedded Retaining Walls as published by the Institution of Civil Engineers insofar as it relates to timber piles, with the addition that faces of piles to receive further fixings shall be aligned to within [1 in 25] (or state other tolerance) of the line shown on the drawings.
    iii) The Contractor shall obtain approval of the proposed piling methods before commencing work.
    iv) The Contractor shall probe for obstructions [at each pile position] [as shown on Dwg. No....] [as directed by the Engineer] [as scheduled ( )]
    v) Obstructions shall be dealt with as instructed by the Engineer.
    vi) Unseated piles shall be re-driven.
    vii) Where there are indications of difficult driving conditions (data to be provided by the Client), the Contractor shall submit proposals for jetting in, pre-boring or phased driving with the piling proposals.
    viii) All pile heads shall be cleaned of burrs, splinters or ragged projections after driving [and damaged rings replaced with new, secured by nails if necessary]
    ix) Pile extensions shall be formed by cutting off the existing pile square and fixing a ....mm x ....mm (type of timber) extension piece ....mm - ...mm

long [complete with pile cap] jointed as Std. Dwg…. [and driven as for main piling] see Dwg. No…. for details

b)   Posts shall be ….mm x….mm (type of timber) ….mm x ….mm long fitted into sockets/post holes/built into ( )] as Dwg. No….

c)   Supports shall be ….mm x ….mm (type of timber) …mm - ….mm long as Dwg. No…. Detail(s)….

d)   Plank piles shall be ….mm x ….mm (type of timber) ….mm long. Piles shall be driven [vertically] [perpendicular to the [bottom] [waling] [shaped] and fixed to the [bottom] waling as Dwg. No…. A margin of [150 mm] (or state other) is allowed above the top of the waling to accommodate the driving head. [Piles in each panel shall be driven 'up-slope']. After driving all piles shall be cut off level with the top of the waling and any splinters or rough edges removed (or as specified, taking into consideration Health and Safety and Public Safety considerations).

e)   Primary Member Constraints.

2.6.   Secondary members

Secondary members provide the main distribution of loads within the structure and are usually fixed by bolting. They include transoms, walings, pier/dolphin/jetty struts, fenders, capping beams, rubbing strakes, kerbs (for vehicle carrying decks) and the like.

a)   (Name of member) ….mm x ….mm (type of timber) [shaped as Dwg. No….] fixed [at ends to (name of receiving member)] [at ….mm centres to (name of member or receiving surface) by (description of fixings) all as Dwg(s). No(s)…. Detail(s) [and jointed as Std. Dwg. No….] [at (preferred joint position(s)].

b)   Secondary Member Constraints.

2.7.   Tertiary members

Tertiary members are generally less substantial timbers, providing cladding and functionality. Examples include groyne planks, deck battens, ramp surfaces, fascias and the like, usually fixed by screws, spikes or dogs.

a)   (Name of member) ….mm x ….mm (type of timber) fixed to [member(s)] with (description of fixings), joints as Std. Dwg. No…. at (description of joint layout) all as Dwg. No…. [Surface treatment (specify)].

b)   Tertiary Member Constraints.

2.8.   Other timbers

This section covers timbers which are not predominantly driven, bolted, screwed, spiked or dogged. It includes timber slotted into steelwork, concrete or other timber and any security fixing i.e. top planks of flood boards, digging in or bolting of bottom planks slotted between king-posts/piles. It also covers unfixed timber (fascines), tied (willow wipes), ground beams, gravity cribwork and earth bank reinforcement.

a)   (Name of member) [....mm x ....mm)] [....mm diameter] (type of timber) [....mm long] [slotted into] [laid on] [tied to] [buried in] (description of member/surface) [and secured by (description)] joints as Std. Dwg. No.... [(joint pattern)] as Dwg. No....

b)   Other Timber Constraints.

2.9   *Site specials*

This section is to contain the variations on the main structures such as end, change of direction, and tie-in details, box-outs, plant bays and the like. It should also contain any architectural embellishments, items likely to be constructed or prefabricated off site, accommodation works, handrails, gates and fencing where they are predominately timber construction. If for some reason one of these elements – say, fencing – becomes a major element of the Contract, it should be listed as a main structure and specified accordingly. Some possibilities are listed below

a)   Plant bay .... planks high as Dwg, No.... [in Groyne(s) No....] [as Layout Dwg(s). No....]

b)   (Name of structure) tie-in to (name of structure) as Dwg. No....

c)   (Name of structure) [end] [change of direction] [box-out] detail as Dwg. No.... [as (list of sites)] [as Dwg(s). No....]

d)   (Architectural detail) [fixed] to (structure and member) [by (type of fixing)] as Dwg. No....

e)   Notice/sign board fixed to [supports bedded in] [structure/surface] by (type of fixing) as Dwg. No.... located [(description of site(s)] [as shown on Dwg. No....

f)   Walkway deck in (type of timber) ....mm x mm [with non-slip surface (describe)] fitted to pontoon as Dwg. No....

g)   Steps consisting of....mm x....mm (type of timber) stringers ....mm - ....mm long with ...mm x....mm treads (type of timber) [and with non-slip finish (describe)] fixed between (head and toe members) [in (name of structure)] as shown in Dwg. No.... Detail(s).... [at (list of sites)] [as shown in Layout Dwg(s). No....]

h)   Handrailing consisting of ....mm x ....mm posts in (type of timber) [fixed to (structure and member)] [embedded in] [at ....mm centres] with (type of timber) [top] railing(s) [shaped as Dwg. No.... Detail...] [and (number) of intermediate railings in ....mm x ....mm (type of timber) at ....centres] [fitted with ....mm x....mm toe board in (type of timber) all as shown in Dwg. No.... Details....

i)   Site Specials Constraints.

Note: - The Dtp. Specification for Highway Works contains Specifications for most of the more commonly used types of timber/timber composite fences and gates. If something different is needed, then follow subsection h) above.

Useful Outline Drawings

Planted Post	Fig. 6.3
Plank Piles	Fig. 6.4
Buried Panels	Fig. 6.5
Props/Struts	Fig. 6.6
Groyne Ties	Fig. 6.7
Overlap Joint	Fig. 6.10
Butt Joint	Fig. 6.11
Scarf Joint	Fig. 6.12
Notched Joint	Fig. 6.13
Double Shear Joint	Fig. 6.14
Coach Screw Fixing	Fig. 6.15
Plate Connection	Fig. 6.16
Steel Angle Connection	Fig. 6.17

Index

abiurana 237
abrasion 40–1, 104, 117, 162, 209
acaria quara 238
access steps 183–5
accidental damage 114
accreditation 88
'actions', limit state design 116
adaptability of groynes 162–4
Afzelia spp. 239
air tools 149
Albizia ferruginea 240
alkaloids 47
amiemfo-samina 240
analyses 7–10, 12–13, 59, 93–4, 103
anchored piles 201–2
ancillaries 123–5
Andaman padauk 241
Andira spp. 242
Androstachys johnsonii 286
angelim vermelho 243
anoxic conditions 196
appraisal process 134–6, 138
Artocarpus spp. 274
Arun District Council 218–19
Aspidosperma spp. 294, 299
assessments 211–13
    degradation 207–11
    environmental impacts 134
    species identification 213–14
    timber structures 207–14
Astronium spp. 288, 312
ayan 244

Bagassa guianensis 305
Balanocarpus heimii 251
balau 245
bank protection 195–9
basralocus 246
baulks 23–5, 131
bed protection 195–9
belian 247
bending strength 39–40
Biocidal Products Directive 61
biological attacks 41–7, 97–100, 207–9
boardwalks 137
boat slipways 183–5
bolts 121, 145, 161

bompagya 248
boron-based preservatives 143
Bournemouth 13–14, 144–5, 217–18
boxed heart 23–4, 64–5
Brachylaena lutchinsii 290
breastworks 168–73
British Standards (BS) 53–5, 57–8, 66
British Waterways 188, 190–1
British Wood Preserving and Damp-
    proofing Association (BWPDA)
    53–5
Bruguiera gymnorrhiza 249
brush box 250
brushwood protection 164–7, 198
BS *see* British Standards
bulkheads 168–9
buried panels 109–10
butt connections 118, 144–5
BWPDA *see* British Wood Preserving and
    Damp-proofing Association

cantilever piles 201–2
capping piles 143
carbon neutrality 6, 76
Caryocar villosum 293
case studies 180–1, 203–5
CCA *see* chromated/copper/arsenate
    preservatives
CDM *see* Construction (Design and
    Management) Regulations
cedars 284
cell types
    hardwoods 20–1
    softwoods 18–19
cellulose 15
certification 84–90, 147
chain of custody 79, 85
chainsaws 150
'checks' (cracking) 178
chemicals
    degradation 209–10
    risks 149
chengal 251
chromated/copper/arsenate (CCA)
    preservatives 49–50, 57–8, 59–60,
    61–2
cladding 200–6

steel
angles 124
composites 68
flat plates 124
straps 124
washers 123
steps for access 183–5
stockpiling procedures 146
stop logs 195
strength 34–40
grading 35–40
identification of species 213–14
species properties 235–6
structural composites 65, 67–8
stresses
member design 113–15
modification factors 37
pile design 153
species properties 37–40
testing 34–5
stress-grading *see* strength-grading
stress-laminating 68
structural size testing 35
structural timber composites 65–71
structure of manual 2–3
structures
construction specifications 313–20
earth retaining 200–6
engineering issues 12–14
enhancements 215–20
existing structures 94–5
life cycle analysis 7–10, 12–13
monitoring and assessment 207–14
specific guidance 155–206
wood 15–18
submerged screens 198
support trestles 181–3
suradan 304
sustainability
definition 79, 81–2
forest management 76–80, 84–6
procurement schedule 91–2
verification 84
swelling 33
Syncarpia glomulifera 307

Tabebuia spp. 267
tar-oil preservatives 51
tatajuba 305
teak 306
technical requirements 82–3, 105–7
Tectona grandis 306

temperate hardwoods 28–9, 157
Teredo spp. 42–4, 46–7, 98–9
tertiary members 112–13, 318
tests
abrasion resistance 41
marine borer resistance 46–7
timber strength 34–5
thatching protection 164–7
thermal degradation 210
tidal currents 101
ties 111, 151
timber
availability 83
case for using 76
choice of structure 96
comparing other materials 9, 11
dogs 123
identification of species 213–14
in the round 57
plates 124
properties 5
selection 105–7, 146–7, 220
working with 133–4
see also species properties
tissue structure of wood 16
tolerances 26–7, 148, 153, 315
tools 149–50, 212–13
toxicity of timber 133–4
tracheids 19
transition wood 45
transverse sections 20–1
Tristania conferata 250
tropical hardwoods
environmental damage 73
groyne timbers 157
procurement 78
readily available species 29
species properties 28–9
strength-grading 36–7, 38
turpentine 307
tyloses 18

uchi 308
UK *see* United Kingdom
ultimate limit state 115–16
United Kingdom (UK)
government policies 77
marine borer survey 99
users of manual 1–2
utile 309